Thomas Henry Potts

**Notes on the Breeding Habits of New Zealand Birds**

Thomas Henry Potts

**Notes on the Breeding Habits of New Zealand Birds**

ISBN/EAN: 9783337144760

Printed in Europe, USA, Canada, Australia, Japan

Cover: Foto ©Andreas Hilbeck / pixelio.de

More available books at **www.hansebooks.com**

# NOTES

ON THE

# BREEDING HABITS

OF

# NEW ZEALAND BIRDS.

## BY T. H. POTTS.

READ BEFORE THE WELLINGTON PHILOSOPHICAL SOCIETY, JULY 17, 1869.

WELLINGTON : JAMES HUGHES, PRINTER, LAMBTON QUAY.

# NOTES

# BREEDING HABITS

OF

# NEW ZEALAND BIRDS.

### BY T. H. POTTS.

(With Illustrations.)

[*Read before the Wellington Philosophical Society, July 17, 1869.*]

#### INTRODUCTION.

THE settlers of New Zealand, so large a proportion of whom are engaged in rural occupations, which placing them in immediate contact with the works of nature, through observation and study ripening into confidential intercourse, will, doubtless, feel deeply indebted to Mr. Buller for his valuable Essay on our Birds, which most interesting division of our Fauna exhibits a notable exception to the comparative dearth of animal life in these islands. When we consider, that from the absence of almost every variety of game, we are debarred the enjoyment of those sporting instincts and habits, which are characteristic of our countrymen wherever they may sojourn, the Natural History of our birds may be found an interesting and useful study, wherewith to beguile many a listless hour; wherever our pioneers of civilization are engaged in subduing the wilderness, near the grateful shade of the forest, in tending flocks on the hill pastures, or cultivating the level acres of the plains.

Nor is it for the merits of that interesting treatise only that we feel thus indebted to its author; its publication has called forth a critical review of it from the pen of Dr. Otto Finsch, of Bremen. The combined result has been of great advantage to the Student of Ornithology, by the removal of certain doubts and difficulties in the nomenclature, and the presentation of a nearly complete list of New Zealand birds, corrected to a recent date. Mr. Buller not only deserves well of his fellow-colonists for what he has done, and merits our thanks for benefits conferred, but we must feel prospectively grateful inasmuch as he is labouring at a complete work on New Zealand Birds. As some time must, however, elapse before his work can be placed in the hands of the public, I beg to offer my small budget of information concerning the mode of nidification and breeding habits of several species, which have come under my own observation, many of which are rapidly becoming scarce. I do so, not only in the hope of its proving of some utility, however slight, but also that others may be induced to communicate their observations, even in a like imperfect shape, and thus lend their assistance in studying our page in the great Book of Nature.

Some of the information here given has been already forwarded to Mr. Buller, at his request, having been gathered from notes and memoranda, made by my sons, and myself, during a long residence in various districts of the Province of Canterbury, where we enjoyed favourable opportunities for pursuing a favourite study.

> "Some to the holly hedge
> Nestling repair, and to the thicket some ;
> Some to the rude protection of the thorn
> Commit their feeble offspring : the cleft tree
> Offers its kind concealment to a few,
> Their food its insects, and its moss their nests.
> Others apart far in the grassy dale,
> Or roughening waste, their humble texture weave.
> But most in woodlands solitudes delight,
> In unfrequented glooms, or shaggy banks,
> Steep, and divided by a babbling brook,
> Whose murmurs soothe them all the livelong day,
> When by kind duty fix'd."—THOMSON.

The Birds of New Zealand present to the observing naturalist most interesting studies in their breeding habits, and various modes of nidification, varying from the compactly-felted nest of the Rhipiduræ, or Mohoua, through easy gradations, every step of which is instructive, till we reach the bare spray-washed rock, on which the Whalebird rears its hardy offspring. They offer to our notice examples of burrowers, troglodytes or semi-burrowers, ground-builders, parasites, and the more or less elaborately-finished structures, which are to be found amongst the incessorial families, in which division the faculty or instinct of bird architecture appears to reach the highest development. Any one who has enjoyed the opportunity, must have been filled with admiration, whilst watching and considering the varying conditions under which the young of different species are reared.* We see that some are fed in the nest until they are well-grown, as the kingfisher and penguin ; others may be said to assist the work of their parents, by following them as soon as they are hatched, and thus materially diminishing the labour of bringing up, by being themselves able to reach the locality of their food supply. Examples of these latter may be found amongst the Rallidæ, Charadriæ, and Anatidæ ; whilst, as observed before, the young of the genera Halcyon and Spheniscus (true burrowing species) remain in their tunnelled holes till well-fledged and well-grown. Yet in the case of *Hymenolaimus malacorhynchus* (which has some claim to be classed as a burrower), a young brood may be noticed with the old birds, on a lake or river, riding on the rippling waves, and floating with buoyancy and ease for hours. The Charadriæ at the best are but indifferent

---

* In a recent work Professor Owen makes this suggestion, "A binary division of the class (Aves) may be founded on the condition of the newly-hatched young, which in some orders are able to run about, and provide food for themselves, the moment they quit the shell (aves precoces) ; whilst in others the young are excluded feeble, naked, blind, and dependent on their parents for support (aves altrices)." See *Anatomy of Vertebrata.*—OWEN.

nest builders, whilst some members of that restless, wary family make no attempt to provide artificial protection for their offspring, the young, warmly clothed with down, appear quite equal to the occasion, and accompany their parents with liveliness and activity. Amongst the Sternidæ and Laridæ, instances may be cited, showing equal indifference in providing shelter for their young; yet, it should be remarked, in these cases the young appear quite incapable of shifting for themselves, and must depend on the industry of the old birds for bringing their food supply to them for several weeks. Here the parental instinct is shown in the selection of the breeding place, the eggs being deposited on the shore of the fishing ground, from whence the food supply of the future family is to be procured; but they have fewer mouths to feed, as they seldom lay more than one or two eggs (*L. Scopulinus, S. Longipennis*). Now, if we turn, for the sake of comparison, to the incessorial genera, denizens of the bush, we shall find the callow nestling equally as helpless as the young in the case of the natatorial birds: but as they number many individuals to each nest, the tax on the energy of the old birds to supply the requisite amount of food, must cause, *pro tanto*, so much the greater labour, unless, indeed, the warmth of numbers closely packed in a well-built nest, should render a somewhat less abundant supply of food sufficient, than would be required to support and rear the exposed broods of the aquatic birds before mentioned.

Some of the Grallatoræ and Anatidæ are remarkable for the extraordinary efforts they make when endeavouring to allure the unwelcome intruder from the immediate neighbourhood of their eggs or young. They will assume lameness, flutter with drooping wing, or drop with a dismal cry into the simulated agony of a death struggle to mislead the wayfarer, but when their artifice has succeeded in enticing him to follow till a safe distance from their precious charge is reached, "they clap their well-fledged wings and bear away," leaving the astonished beholder to meditate on the folly of trusting to appearances.

Amongst the troglodytal birds (such as Nestor, Platycercus, etc.) white is the usual colour of the eggs, doubtless as a provision to render their position more readily seen in the dim twilight of the breeding place, consequently to be approached and incubated with less danger of accident. It is, however, amongst the ground-breeders that the colouring of the eggs elicits the admiration of the careful observer; in some instances (such as Anarynchus frontalis) so wonderfully does the colouring of the eggs harmonize with the prevailing tone of the ground on which they are deposited, that accident only would disclose their presence to the casual wanderer, besides the instance just given, *H. Longirostris, L. Scopulinus*, afford noteworthy examples of this peculiar property which offers such a reliable safeguard against the plundering propensities of bipeds, whether feathered or not.

The rapid diminution in the numbers of our birds (with the exception of a very few varieties, of which *Zosterops lateralis* is the most noticeable instance) must be apparent to everyone who has given the slightest consideration to the subject, it is a matter of deep regret that, in all probability, many species will have become extinct ere their habits can be sufficiently studied by the naturalist for their use, economy, and position in our Fauna to be correctly ascertained. To the future student of the natural history of our country, vague, unreliable traditions, a conflicting nomenclature, and the contorted productions of the taxidermist mounted in acrobatic and weird-like attitudes, will perhaps alone remain to fill up the hiatus. How far should we now have to travel to discover a bevy of Quail, in the seclusion of some very remote valley of the "back country," a straggler or two might be met with. Yet by referring to the "New Zealand Handbook," it may be noted that the large island in Port Cooper was named after this bird, from the number of Quail flushed there. The beautiful little Rails are now almost as scarce; in how few streams or lakes could one now expect a sight of the noble White Crane, watching " with motionless regard," its finny prey. yet but twelve years since, the writer of this paper gave Lake Heron, in the Ashburton country, its name, from the numbers of this majestic bird, which frequented its shores, or soared above its surface with lazy, heavy flight. These are but a few names of rare birds from a list that is annually increasing; and it is a matter of no great difficulty to point to the causes, which have led to what must certainly be deemed a misfortune to the Colony.

As the country became occupied, the more remote districts rendered accessible by means of roads, as wide-spreading swamps were drained and brought into cultivation, extensive tracts of country stocked with cattle and sheep, above all, as the whole face of the country became changed by the repeated bush fires, it can be readily understood how these various incidents of civilization should so soon have effected such considerable changes in the condition of our feathered tribes. To these other minor causes may be added, and, perhaps, contemplated with less satisfaction, the reckless gunner frequently killing for the mere love of slaughter, the self-complacent "new chum," with the inevitable firearms, even the learned savant will sometimes be tempted to destroy both old and young, especially of our rarer birds, a favourable opportunity of procuring choice and desirable specimens being too great for resistance; scientific zeal thus overcomes good policy, and consideration for the future. Would not the re-establishment of some of our rarer species (many of which are excellent as articles of food) form a worthy, if not a noble, object of ambition for our Acclimatization Societies to achieve? * The difficulties attending such an undertaking would necessarily be great, there is no doubt, but difficulties energetically encountered are seldom insuperable. To ensure anything like a successful issue being arrived at, certain conditions must be fulfilled, early action must be taken, an intelligent acquaintance with the habits of birds, would be indispensable, patience to endure considerable trouble, as well as occasional disappointment, and lastly, a small expenditure of money. However, a moderate outlay should not be an impediment to an undertaking of such interest with those institutions, which have been handsomely supported by private subscriptions, supplemented by liberal grants of public money.† The Parliament of New Zealand has taken steps to protect some of our birds, but however well legislative enactments may be framed, the people themselves can alone determine what shall be allowed to exist; looking at the rapid destruction threatening our noble forests, and in some cases our fisheries also, it must be admitted that the prospect of the preservation of our birds is the reverse of hopeful.

As a rule, we appear to live, work, and legislate for to-day, with not too much anxious thought for the to-morrow of those who are destined to succeed us. Whatever may have been the result in other countries which have been colonized by our race, whoever fairly writes the history of this country, will have to record how deeply the Anglo-Saxon settler has implanted his mark, by the alteration of the natural features it presented on his first arrival. Perhaps it would not be considered out of place to offer a few general, but very brief,

---

* "One of the exciting causes of the destruction of every living native animal that can be met with is the pretence of enriching our Museums, while at the same time the overstocked market in Europe render them, for the most part, unsaleable there; and it is a well-known fact, that the skins of Australian birds, etc., have been re-exported from England to Australia for sale."—See Dr. Bennett's "Gatherings of a Naturalist in Australia."

† Should our Acclimatizing Institutions require a precedent, they may refer to the "Bulletin de la Société Impériale Zoologique d' Acclimatation, 1864." Among the extraordinary prizes offered by the Imperial Society, February, 1864, may be found,—Reproduction in France of the Pinnated Grouse (*Tetrao Cupido*) la Gelinotte, medal of 1,000 francs.

remarks on the study of our ornithology, which presents a wide field for the instructive entertainment, even of those who do not enjoy the advantage of much out-of-door occupation, as diligent investigation will repay those who are disposed to devote time and attention to its careful consideration. The tegumentary system of birds is so remarkable and beautiful a feature, amongst the glories of Nature's handiwork, as at once to call for observation, the most heedless must be attracted by the exquisite arrangement of form and colour. Although man's chief interest in the feathered tribes centres, in the fact of their furnishing him with delicate and luxurious varieties of food, or amusement, and a mild excitement in the way of sport; yet several species are so lovely in their gorgeous trappings, that birds of many kinds are kept in a semi-domestic state, merely for the gratification their beauty imparts to the beholder. Vast numbers, more fortunate perhaps, are eagerly pursued and slain, not for economical purposes of supplying food or clothing, but that their rifled plumes may be worn as articles of personal adornment. Human vanity has long since established this custom so universally that neither age, sex or race appears exempt, and the chief of a Maori tribe doubtless feels as much pride in his feathered head-dress plucked from the beautiful train of the snow-white Kotuku, as the bedizened wearer of ostrich plumes, whether a prince or a peeress of one of the most civilized nations of Europe.

The Parroquet may be frequently observed in confinement, and the Tui, liveliest of our Meliphagidæ, quite as often perhaps barred within the limits of a dirty cage, has to exchange, for the dewy nectar of fresh bush flowers, a monotonous diet of soaked bread or biscuit, and for this unpalateable dole the unfortunate prisoner for life is expected to be lively and gay. The Maories of the South Island have long prepared the Mutton-bird, as a dainty article of food.

What can be more wonderful than the development from the inert contents of an egg, of so sprightly a creature as a bird; if we were not familiarized with this admirable and curious process of nature, it would be deemed miraculous; it really appears typical of the Creation, and this must have been felt, where the egg was looked upon as the symbol of the renovation of the living world, and the custom was introduced, of suspending an egg in Eastern Churches. A writer on the monasteries of the East says, "as the egg contains the elements of life, it was thought to be an emblem of the ark, in which were preserved the rudiments of the future world."

Passing over the embryological age, the period of incubation which represents the term of gestation amongst mammals, and the growth of the young in all its stages of dependence, our attention is arrested by the anatomical structure of this class of vertebrates. The peculiar arrangement of the osseous and muscular systems, from whence the powers of locomotion in all their admirable variety are derived, should be carefully considered, flying, walking, hopping, climbing, swimming, or diving, from the constant exercise of which, birds depend for safety, or obtain their food supply.

A transient glance at the structure of their skulls and beaks will satisfy the enquirer how happily their forms are adapted for the habits and varying conditions of the life of different species. The strong hooked beak of Nestor, by the help of which it rapidly ascends the stems or branches of trees, is sufficiently powerful to rend down long strips of tough bark, such as that of *Fagus solandri*; the soft bill of Hymenolaimus enables it to secure small aquatic insects, caddis worms, etc., in the mountain creek; the slender curved mandibles of Recurvirostra are fitted for thrusting into the oozy slime of the swampy marsh; with its strong beak, the cosmopolitan Hæmatopus readily breaks the shell-armour of the various bivalves that pave the tide-washed mud flats of our harbours; the reason for the lateral curvature of the beak of the

Anarynchus. or Crook-billed Plover, as yet requires explanation, which can only be satisfactorily given from a closer observation of its habits. Compare the bones of Himantopus, the graceful wader, with those of Podiceps, or Sphœniscus, chief amongst divers, the wabbling, yet undulating gait of the latter, when hastening to the sea, gives it rather the appearance of a large water-rat than that of a bird, but how its awkwardness on shore is compensated for, anyone may judge who witnesses the ease and rapidity with which it dives beneath the swelling wave, by the aid of its fin-like wings. Notably, Fregata, Diomedia. Thalassidroma, with their enormous development of the bones and muscles of the wings, their consequent almost untiring flight,* offer the most remarkable contrast to many species, such as Apteryx, Ocydromus, and the wingless giant peculiar to our land, which perhaps has not been long extinct. It is interesting to note that it was from a bone of this genus that *the* instance of the perfection of skill to which the accomplished anatomist can attain, was exhibited, as all the world knows, by Professor Owen building up, from the study of a single bone, his theory of the gigantic bird, the correctness of which was afterwards corroborated by the discovery of ample remains of various species of Dinornis. Is it possible that the Moa was known to the ancient world? The following passage from Strabo would answer for a description of its pursuit by natives, quite as well as for the hunting of the Dodo of the Mauritius, or the Æpyornis of Madagascar. Writing of the countries washed by the Red Sea (Book xvi.), Strabo observes, "Above this nation is situated a small tribe, the Struthophagi (or bird-eaters), in whose country are birds the size of deer, which are unable to fly, but run with the swiftness of the ostrich. Some hunt them with bows and arrows, others covered with the skins of birds, they hide the right hand in the neck of the skin, and move it as the birds move their necks. With the left hand they scatter grain from a bag suspended to the side ; they thus entice the birds till they drive them into pits, where the hunters dispatch them with cudgels. The skins are used both as clothes and as covering for beds." Such an ancient notice of a wingless bird is interesting.

The flight, migration, sight, and voice, of many of our species of birds, are all subjects of interest to those who are glad to learn something more of the world we live in.

When the Lark is flushed from her nest on the wide expanse of the tussock-covered plains, with what rare instinct or wonderful gift of sight must she be endowed, which enables her to find her nest amidst the myriads of tussocks presenting the same aspect, without a track, a tree, or even a rock, as a guide to aid her organ of locality. How true is the Bronze-winged Cuckoo to his appointment, almost to a day, the first week in October he announces, by his presence, that high spring has been reached, and the active labours of our portion of animated nature has commenced in earnest.

We cannot boast of possessing, amidst our bushes, rivals to those "melodious songsters of the grove" which wake up the woods and hedgerows of the Old Country, yet many of the notes and cries of our feathered race are peculiarly interesting, such as the song of the *Petroica albifrons*, the human-like whistle of the *Chrysococcyx lucidus*, the well-known chime of the Bellbird, the extraordinary sounds to which the white banded Tui gives utterance, the flute-like tones of the Crow or Wattle bird, the wailing call of the Weka; and the startling shriek of that night bird, frequently heard in the back country, which has not been identified as the call of any bird that has yet been described.

---

* After the memorable storms of July and August, 1867, in Lyall's Bay, amongst numbers of Hapuka and other fish that had been stranded, we observed several bodies of *Diomedea exulans*, that had perhaps been dashed against the rocky cliffs, by the violence of the storm.

For years attempts have been made to procure a specimen of this mysterious unknown, which will probably be found to belong to the families either of Strix or Podargus; it is to be hoped it may not turn out to be the man-liking bird thus mentioned by Fuller, " I have read of a bird which hath a face like, and yet will prey upon, a man, who coming to the water to drink, and finding there, by reflection, that he had killed one like himself, pineth away by degrees, and never afterward enjoyeth itself."

Already some of our rural settlers attach significance to the peculiar flight and cries of birds, as prognosticating changes in the weather, thus fol-lowing out in their new home the like fancies or observations which have been handed down by their fathers from time immemorial; on this subject Cuvier wrote, " For the rest of their intellectual qualities, their rapid passage through the different regions of the air, and the lively and continued action of this element upon them, *enables them to anticipate the variations of the atmosphere,* *in a manner of which we have no idea, and from which, has been attributed* to them from all antiquity, by superstition, the power of announcing future events."

Embryological research as far as our birds are concerned is still a sealed book. This is a branch of science upon the importance of which Agassiz lays much stress; after speaking of the information he had acquired from the examination of bird embryos, he writes, " How very interesting it will be to continue this investigation among the tropical birds!—to see whether, for instance, the Toucan, with its gigantic bill, has, at a certain age, a bill like that of all other birds; whether the Spoonbill Ibis has, at the same age, nothing characteristic in the shape of its bill. No living naturalist could now tell you one word about all this." Investigations of this nature amongst the several genera peculiar to New Zealand, would be of value to science, and would offer an interesting field for new discoveries concerning ornithological facts, in our bright corner of the world, which the scientific naturalist has not yet found time or opportunity to lay bare.

Accuracy of description is so necessary to establish facts, that it is far preferable to give a few brief notes, the result of actual observation, rather than to supply pages of information gathered from hearsay; even in our humble researches, the untrustworthy character of report generally, has been experienced sufficiently often, to impart a certain amount of incredulity not easily shaken off; mythic treasures have so frequently eluded pursuit, when the scene has been reached that should have disclosed specimens of more than ordinary interest, that no difficulty is felt in understanding how often fable creeps in, and becomes, in a measure, blended with truth in matters relating to Natural History.

On the other hand it is far from safe to discard even the improbable, as imperfect description has before now converted the improbable into the apparently impossible, as a very early notice of the Hornbill will testify.*

* In 1330, Odoric tells of a bird as big as a goose, with *two heads.* In 1672, P. Vin-cenzo Maria describes a bird, also as big as a goose, but with *two beaks,* the two being perfectly distinct, one going up and the other down; with the upper one he crows or croaks, with the lower he feeds, etc.— *Viaggio,* p. 401.

In 1796, Padre Paolino, who is usually more accurate, retrogrades; for he calls the bird "as big as an Ostrich." According to him, this bird, living on high mountains where water is scarce, has the second beak as a reservoir for a supply of that element. He says the Portuguese call it l'assaro di duos bicos.— *Viag.,* p. 153.

Lastly, Lieut. Charles White describes the same bird in the Asiatic Researches. " It has a large double beak, or a large beak surmounted by a horn-like shaped mandible."— *Asiatic Res.,* iv., 401. The bird is a Hornbill, of which there are various species having casques or protuberances on the top of the bill, the office of which does not appear to be ascertained. How easy here to call Odoric a liar! but how unjust, when the matter has been explained.— *Cathay and the way thither,* Vol. i., p. 100.

Many writers of Natural History appear to have made a practice of copying from their predecessors: the inconvenience of this arrangement is manifest, in that errors were thus allowed a very protracted existence, such as the fables which were for centuries supposed to describe the natural habits of the Kingfisher, etc. The writer of this paper was long haunted by the vignette title of a popular work on British Birds, the engraving was supposed to give a correct representation of *Cinclus aquaticus*, and nest; the latter as there figured, presented the conventional basin-shaped arrangement with eggs, all complete, the popular notion of a bird's nest in fact; now, in reality, the nest is a thick mossy dome-shaped structure, in which the pure-white eggs are concealed from view. Years after quite as great a shock was felt, when on inspecting a public collection, he found that if he placed reliance on what he saw before him, Falcons must have laid Pigeon's eggs, Seagulls had produced those of the Turkey, whilst the Crested Grebe had achieved a Duck's egg. Careless mystifications such as these, should be avoided by those who are expected to impart information, as *too improbable*.

An attempt to show, more clearly, the extent which the ravages of a few years have inflicted on the numbers of our birds, may perhaps be excused for the object in view, we will therefore endeavour even at the risk of being tedious, to represent such a scene of the past as one might reasonably expect to meet with, almost daily, during a considerable portion of the year, at the place indicated. One of the most favourable localities for observing the habits, acquiring a knowledge of the notes and cries, and watching the flight of various birds, was not far from the gorge of one of our great southern rivers, where the monotonous flatness of "the plains" gives way to a more broken and undulating surface, as an extensive range of hills is approached. This range is on one side flanked by low downs enclosing a few shallow lagoons, here and there sparsely-wooded gullies intersect the hills, from whence flow two or three brawling creeks, that join and deepen into a swift and silent stream crossing the grassy flat; the higher portion of this corner of "the plains" is stoney, whilst near the foot of the downs lies a swamp of no great extent.

Here upwards of thirty varieties of birds might be observed almost daily, and here too, or within a very moderate circuit, most of them breed.

Then our handsome Quail abounded, flying straight and low when flushed; the finding its slight humble nest filled with eggs, was no rare occurrence; or to see from amidst the snow-grass tussock, the Weka confidently emerge, or to hear the little Grass-bird utter its unchanging note u-tick, u-tick, as rising on feeble wings that just sustained it to the sheltering grass, beneath the spreading leaves of a neighbouring flax bush, whence perhaps the Tit (Petroica) darted to the ground from the tall flower-stalk, to snatch the larvæ of the grasshopper. Then the blue Pukeko, prince of Rails, often stalked through the raupo of the swamp, or the brown-streaked Bittern, with long ruffled neck, rose with deliberate flight; perchance hard by in the narrow outlet bounded by tufted stumps of carex, the light-eyed Teal slunk silently from view; or further on, where the creek widened to a noiseless pool, the little Grebe with rosy breast, dived and sported with restless activity; close by a group of sober Grey Ducks; whilst the watchful Paradise Drake basked on the sunny bank above, his beady eyes doubtless commanding a view of a certain snow-grass tussock, under the waving plumes of which, a cup-like nest of down might lie securely hid. Then perhaps amongst the tall feather-tufted tohe-tohe reeds, and saw-edged grass, a pair of Harriers had built their rough, flat-topped home, or floating high above on noiseless wing, alarmed the pyeball Redbill, that circles round on rapid wing, screeching its clamorous note; or we might watch the pied Stilt with long pink legs, outstretched rudder-like behind, making for the rush-fringed lagoon, to join its mates in wading near the margin of the pool, whose placid surface, now broken into a thousand ripples, as it shivers beneath the touch of the passing breeze, laden with sweet perfume, collected from the thorny Discaria, the formal solitary Cordyline, or the creamy bells of the brown-leaved Epacris. Now perhaps behind a favouring flax bush, we watch the visitors that dot the surface of the water (amongst them, the Black Widgeon and variegated Shoveller were rarely to be seen) and observe some early flappers skimming along in hot pursuit of their insect prey. Crossing towards the higher stony ground over patches of gizzard-stones, and many a bleached bone, crumbling in decay, of the giant Moa, that tells a tale of days philosophers may dream of; perhaps the sprightly lark, with lively chirrup, mounts from its freckled eggs, or the banded Dotterel flies round with warning note, whilst its grey-clad young hide cunningly behind some stick or stone; or red-billed Terns gather round in screaming flocks, returning from a blackened patch of new-burnt ground, that stretches far out on the plains, whilst from many a beak dangles the writhing lizard; or maybe the slowly repeated twit, twit, of the red-breasted Plover chimes in, as it sidles slyly off with alternate run and halt, nor could you find its slight grassy nest till half a dozen times the ground had been stepped over. The rock-bound gully reached (the heights above, as New Year's day came round, ablaze with crimson Rata flowers), from the swift stream below, amidst its noisy brawling with the rocks, arose the plaintive whistle of the Blue Duck, as with soft-fringed bill it explored each little foaming eddy; or scrambling through the scrub, we might observe, on the rifted top of a huge lifeless tree, the great Black Shag, perched motionless; beneath, Bell-birds, with noisy blustering flutter, seek the konini, clinging to its brittle sprays, extract the honey of the pendant flowers; or high up, clear into the golden glow of sunshine, ascends the glistening Tui, discharging a whole volley of strange sounds; or perhaps from the rocky bush, the green-clad Parroquet descends, its harsh note repeated rapidly; where sand-flies gather thickest and irritate the rambler with their dusky swarms, the Fly-catchers, pied and black, flit around, then perching, spread their fan-like tails with twittering chatter, whilst from a bare branch above, the strong-billed Kingfisher keeps watch above the gurgling creek. Then we might note where the small striped Wren crept round the lichen-covered trunk, or moss-clothed branches of some spreading shrub, or the grey warbler (Piripiri) with quivering notes fluttered near its cosy, dome-shaped nest; perhaps on a huge blackbirch the Kaka might be seen rending down the bark in long ribbon strips, to reach the insect dainties that lay housed beneath; or, with rapid flapping wing, the Pigeon seeking the straight-stemmed Kohi, whilst concealed by the rising tiers of leafy canopy, the bronze-winged Cuckoo whistled from the topmost bough. Emerging from the bush's dusky light, into the full glare of noon, we might perhaps have seen the Quail-hawk, rapidly ascending with spiral flight, till it appeared like a dark speck against the cloudless sky, its shrill jarring scream distinctly heard the while. Descending through groves of formal Ti palms, the steep, stone-paved terraces of the great river that rushes in milky streams below, the large Grey Gull might perhaps be found feasting on the carcase of a sheep, stretched on a patch of dark-green tutu; or hard by the margin of the sandy spit, the little Gull was perched neat and trim as any quakeress, whilst the Black Stilt, with its uneasy cry of pink, pink, settled a few yards onwards, to lead us from its crouching young, or the Crook-billed Plover scuttled slowly off with outstretched wing. Those less common birds, the great White Crane, Avocet, and Spoonbill Duck were seen at rarer intervals.

Now the scene is changed, and so thoroughly; it seems almost like a dream that such things were. The wooded gulleys denuded of timber, show amidst blackened stumps, some isolated shrubs, still green, of olearea, panax, or much-enduring coprosma; the constantly recurring bush fires have cleared off the stately Ti palms (so fragrant in early spring); dwarfed flax bushes, altered the condition of various grasses, improving some for grazing, effected a speedier drainage, and dried up the shallow lagoons. Thousands of sheep now depasture on that well-remembered corner of "the plains," on those gently-swelling downs; instead of the varied cries of birds we have the bleating of flocks, the bark of the colley as it *rounds up* its charge, the loud crack of the stockwhip, the hearty curse of the bullock driver delivered "ore rotundo;" these changes form part of the evidence that testifies to the progress of our civilization.

If from some of the causes thus pointed out, or the rapid rate at which the timber forests have been wasted or destroyed,* the introduction of bees (and the numbers of swarms met with in the bush may easily account for some diminution in the food of the Meliphagidæ), the spread of cats, and even rats, or from the feeble hold on life which appears to be shared by every living thing that is indigenous, whether animal or vegetable, when brought into contact with foreign influences, it should be deemed impossible to avert the impending fate which threatens the existence of many species of our native birds, we must endeavour to find some compensation for so great a misfortune, in the success which has attended the introduction of foreign birds in many parts of the country. The Pheasant, Partridge, and Californian Quail, are amongst the best of the game birds that may be considered as established amongst us. The Black Swan, introduced in Canterbury to check the growth of another foreigner (watercress), Shell Parroquet, Thrushes, Blackbirds, Larks, Chaffinches, Greenfinches, Sparrows, Starlings, etc., from increasing numbers, promise very soon to give additional interest to our rural scenery.

---

## LIST OF BIRDS

DESCRIBED IN THIS PAPER, WITH THE MEASUREMENTS OF THEIR EGGS.

| No. | | Length. in. | Length. lines. | Breadth. in. | Breadth. lines. |
|---|---|---|---|---|---|
| 1. | Falco Novæ Zelandiæ, Gml. | 2 | 0 | 1 | 6 |
| 2. | Circus assimilis, Jard. | 1 | 11 | 1 | 6 |
| 7. | Halcyon vagans, Gray | 1 | 0½ | 0 | 10½ |
| 10. | Prosthemadera Novæ Zelandiæ, Gml. | 1 | 2 | 0 | 10 |
| 11. | Anthornis melanura, Sparrm. | 0 | 11 | 0 | 8½ |
| 15. | Pogonornis cincta, Dubus. | 0 | 9½ | 0 | 7 |
| 18. | Acanthisitta chloris, Sparrm. | 0 | 7¼ | 0 | 6 |
| 19. | Mohoua ocrocephala, Gml. | 0 | 10½ | 0 | 8 |
| 20. | Sphenœacus punctatus, Quoy. | 0 | 10 | 0 | 7¾ |
| 25. | Gerygone assimilis, Buller | 0 | 8 | 0 | 6 |
| 26. | Certhiparus Novæ Zelandiæ, Gml. | | | | |
| 27. | „ albicilla, Less. | 0 | 10½ | 0 | 7½ |
| 29. | Petroica macrocephala, Gml. | 0 | 9 | 0 | 7 |
| 31. | „ toi toi, Less. and Garn. | 0 | 9 | 0 | 7 |
| 32. | „ longipes, Less. and Garn. | | | | |
| 33. | „ albifrons, Gml. | 1 | 0 | 0 | 9 |
| 34. | Anthus Novæ Zelandiæ, Gml. | 0 | 10½ | 0 | 8 |
| 35. | Zosterops lateralis, Lath. | 0 | 8 | 0 | 6½ |

---

*According to a return recently laid before the Provincial Council, over upwards of 170,000 acres of bush land, have depasturing licenses been granted by the Waste Lands Board of the Province of Canterbury. Is it the interest of the licensees to preserve timber?

|  | in. | lines. | in. | lines. |
|---|---|---|---|---|
| 37. Rhipidura flabellifera, Gml. . | 0 | 8 | 0 | 6 |
| 38. „ fuliginosa, Sparrm. . | 0 | 8 | 0 | 6 |
| 47. Platycercus Novæ Zelandiæ, Sparrm. | 1 | 1½ | 0 | 10 |
| 50. „ auriceps, Kuhl. . . | 0 | 11½ | 0 | 9½ |
| 51. Nestor meridionalis, Gml. . | 1 | 9 | 1 | 3½ |
| 58. Chrysococcyx lucidus, Gml. . | 0 | 9 | 0 | 6 |
| 60. Coturnix Novæ Zelandiæ, Quoy. | 1 | 3 | 0 | 11 |
| 61. Apteryx australis, Shaw . . | 5 | 1 | 3 | 4 |
| 62. „ Oweni, Gould . . | 4 | 6 | 2 | 7 |
| 63. „ Mantelli, Bartl. | 5 | 4 | 3 | 3 |
| 65. Charadrius bicinctus . | 1 | 4 | 1 | 0 |
| A. 65. „ obscurus, Gml. . | 1 | 9 | 1 | 3 |
| B. 65. Anarhynchus frontalis, Quoy. . . | 1 | 4½ | 1 | 0½ |
| 71. Hæmatopus longirostris, Vieil. | 2 | 3 | 1 | 7½ |
| 75. Botaurus poicilopterus, Wagl. . | 2 | 1½ | 1 | 6 |
| 78. Himantopus Novæ Zelandiæ, Gould | 1 | 10 | 1 | 3 |
| B. 78. „ melas, Homb. . . | 1 | 10 | 1 | 3 |
| 87. Ocydromus australis, Sparrm. | 2 | 2½ | 1 | 5½ |
| 91. Porphyrio melanotus, Temm. . | 2 | 0 | 1 | 5½ |
| 92. Casarca variegata, Gml. . | 2 | 7 | 1 | 10 |
| 93. Anas superciliosa. Gml. . . | 2 | 3 | 1 | 9 |
| 94. „ chlorotis, Gray . . | 2 | 5 | 1 | 10 |
| 96. Fuligula Novæ Zelandiæ, Gml. . | 2 | 8 | 1 | 9 |
| 98. Hymenolaimus melacorhynchus, Gml. | 2 | 8½ | 1 | 9 |
| 99. Podiceps rufipectus, Gray . . | 1 | 9 | 1 | 0 |
| 100. „ Hectori, Buller . . | 2 | 4 | 1 | 7 |
| 104. Spheniscus minor, Forst. . . | 2 | 3 | 1 | 9 |
| 126. Larus Dominicanus, Licht. . | 2 | 10 | 1 | 10 |
| 127. „ scopulinus, Forst. . . | 2 | 1 | 1 | 6 |
| 129. Sterna caspia, Pall. . . | 2 | 7 | 1 | 9 |
| 130. „ longipennis, Nordm. | 1 | 10 | 1 | 4 |
| 131. „ antarctica, Forst. . | 1 | 6 | 1 | 1½ |
| A. 131. „ sp. (Sternula nereis), Qy. | 1 | 4 | 0 | 11 |
| 139. Graculus brevirostris, Gould | 2 | 6 | 1 | 6½ |
| 142. Dysporus serrator, Banks . . | 3 | 1½ | 1 | 10 |

It may be interesting to persons acquainted with the Oology of Europe, to institute a brief comparison between the eggs of some of our birds, and those of kindred European species; in some few, considerable contrast in size and shape, may be observed; whilst amongst others so little difference is to be discerned, that it would be difficult to decide, from transient inspection, of which hemisphere they are native.

The eggs of *Falco Novæ Zelandiæ* closely resemble those of *F. peregrinus*, in size, form, and colour; so also do those of *Circus assimilis* bear as striking a likeness to those of *C. rufus*. The eggs of *Halcyon vagans* are larger than those of *Alcedo ispida*, the same may be said of those of *Coturnix Novæ Zelandiæ*, when compared with those of *C. vulgaris*. To select the eggs of *Hæmatopus longirostris*, from a number of those of *H. Ostralegus*, would be difficult; nor would it be much less so to decide whether the Bittern's eggs were European or New Zealand; the eggs of *Himantopus melanopterus* strongly resemble those of our Stilts, the same remark will apply to those of *Podiceps minor* and *rufipectus*, respectively. With regard to the eggs of *P. cristatus*, they are smaller than those of *P. Hectori*. The eggs of *Sterna caspia* bear a very close resemblance in both hemispheres. The similarity between the eggs of *Sterna minutæ*, and the new species from the Rakaia, has already been pointed out. The egg of *Dysporlus serrator* only differs by 1½ lines in length, from that of *Sula alba* of Europe ; whilst similar chalky encrustations may be found on either specimen.

## No. 1.—FALCO NOVÆ ZELANDIÆ, Gml.
### Ka rewa rewa-tara.
### Quail-hawk.

In New Zealand, the courageous family of the Raptores is very feebly represented, the honourable post, of head of the family must fairly be assigned to this bird, which is commonly known by the name of the Quail or Sparrow-hawk ; " the hardy Sperhauke eke the Quales foe." as Chaucer has it. This bold little Falcon, which, a few years since, was so frequently seen, is now of comparatively rare occurrence. How seldom do we now hear that wild chattering scream, which gave notice of its approach, and spread alarm amongst the denizens of the poultry yard. Many instances might be cited of its daring courage and perseverance in pursuit of its prey, such as dashing into houses, penetrating to an inner room, striking its quarry, and clinging to it till ruthlessly knocked over with a stick. Years ago, when Quail shooting, how we have been troubled by the assiduous attendance of this bird, and have shot this dauntless fowler almost in the act of swooping off our game. We have noticed the female, with a Tui trussed in her talons, which she carried a considerable distance without a rest, when the male soared boldly in company, and kept watch and ward over his well-laden helpmate.

At present it is in the "back country" only, that we can hope to find its breeding-place, which is usually on a ledge of rock commanding a prospect over some extent of country. Such an out-look gives an advantage of no little value, of which the Falcon is not slow to avail itself, should such a bird as a Tui or Pigeon appear in sight.

Several of the breeding-places, which we have had opportunities of examining, have presented, in a remarkable degree, very similar conditions as regards situation. Amongst bold rocks on the mountain side, somewhat sheltered by a projecting or overhanging mass, appears to be the favourite site for rearing its young. The eggs very closely resemble those of *Falco peregrinus* of Europe, in colour, size, and shape, usually three in number, are deposited on any decayed vegetable matter, that wind or rain may have collected on the rocky ledge, for the efforts of this bird in the way of nest building are of the feeblest description. The eggs are of a rich reddish-brown, mottled over with darker shades of brown, sometimes the ground-colour is pale reddish-white, less suffused with the darker colour at the smaller end, broadly oval in shape, they measure 2 inches in length, with a diameter of 1 inch 6 lines. Some eggs taken from a range near the head-waters of the Rakaia, give measurements somewhat less than the above, with a yellowish, in place of reddish-brown colour. Young birds are covered with grey down at first, and assume a plumage of dark brown above, with breast of rufous-white spotted with brown, thighs slightly rufous. October, November, and December is the principal breeding season, and the localities we have noted for the eyries, are rocks near Cass's Peak, Governor's Bay, Malvern Hills, River Potts, Mount Harper, etc.

NOTES.—Oct. 10—Young Quail-hawks, near the home paddocks on the Rangitata River.

Nov. 8—Above the upper gorge of the Ashburton or Hakctere River, found a nesting-place on the bare soil, sheltered by a large isolated rock ; two young birds, covered with grey down, old birds very bold in defence of their young.

2 Circus assimilis
   H. novæ...? Fi...

(approximans)
(Gouldii)

4. Athene albifacies
7. Halcyon nagaus
11. Anthornis melanu...
12.  "  melanocep...
       brunin gyf...
14.a ~~Mohoua albirita~~
       .....
22. Gerygone igata :
    "  (flaviventris
    "  (assimilis)
24.  "  albofrontata
34. Anthus nov.-zeel.
37. Rhipidura flabe...
45. Creadion carunc...
55. Stringops habropt...
57. Eudynamis laiten...
58. Eurystomus lucidus
       ...
59 Carpophaga nov...
    purple sports on...
61 Apteryx australis
63  "  (Mantelli
       fu... gl...
62  "   (Oweni
65. Charadr. bicincta
       ...
       ...
66  "  ga...
    (seebohm - ...
? Thinornis (Hæmat...?
    ...
    ...
    ...

Hæmatop. longirostris
       ...
    "  unicolor
       ...
Ardea intermedia     425-5...
   "       "        48.5...
   "     sacra       40.2-...
   "       "
Botaurus pæcilopterus
       ...(Camp...
Nycticor. caledonicus
    "   flavipes
    "       "
Himantop. nov. zeel.
Limosa uropygialis

Recurvirostra rubricollis
    ...
    ...
Rallus assimilis Gr. (...
       ...
Ortygometra tabuensis
       ...
Ocydromus australis ...
       ...
Porphyrio melanotus M. 50
    ...
    "       "      53.2 +
Anas superciliosa  59  5
   "      "         59
   "      "         57
Podiceps Hectori   60.3
Aptenodytes patagonicus 10...
    "       "         103

the
ften

Creur assimilis. 39.2 + 32.6, ...lich roform., ...ß ...w. mit ...
H. ...? Finch, 4 v. Palau-I.

48.5 + 38 ...ß (Goald, B.A.)
44.2 + 35.4, ...ß, mit ...gen ...ßärmen. A. /...ten, ...
(approximans) 42 + 34 bläulißmaiß /...ß. (Tienen.)
(Gouldii) 47.2 + 35.5 mit ...ßen ...güm. ...Walb. (...)

7 Athene albifacies 46.3 + 70, ...ß, ...ßh... 1.0.4? Ava, 84. ...
9 Halcyon vagans 27.3 + 22.8, ...ß, (Hutton, Ibis 70)
11. Certhiparus melanura 21.4 + 16.4 ...ß, ...ge... (Hutten ...
12 " melanocephala 26.5 + 18.9 ...ßl. ...ler ...ß als ..., ...f...
... gaß. (Travers, Ibis 72)

17 Mohoua albicilla, ... Potts ...ung f. ... Wellington ... ...
... ... H. albicephala = 29 Certhipar. albicilla.

22. Gerygone igata:
" (flaviventris) 13 ...ger gro ... 3. f. w. assimilis und ganz ...ß /...
" (assimilis) bianform...: ...f. ... ...ß. ...ü... u. ...tl. Ll. ...
24. " albofrontata 18.7 + 13.6, ...ß. ...ß, u. ... ... ...
34. Anthus nov.-zel. 22.8 + 17.7, gelblißgrau ...ün ... ... (Hutten
37. Rhipidura flabellifera 16.4 + 12.6, ...ß, ...ün. für Kl. ... Ll. ...
45. Creadion carunculata 29.3 + 22.8 (bläulich /...ß, ...ün ...
55. Stringops habroptilus 54 + 39.6, ...ß (Goald, B.A. /Kayflang)
57. Eudynamis taitensis ...ß. ...äul. ...ße ... für ... Nehm. J. 79, ...
58. Eurystomus lividus 17.4 + 12.7 ...dunkl..., an der ...ße ...er (...
... ... ...r ... ... ...ß...
...lich.
59 Carpophaga nov. zel. 35.4 - 37.2 + 27.5; ...ß v. ...ß, ... mit ...
purple spots on the larger end" (Travers l.c.)

61 Apteryx australis 129.1 - 185.4 + 82.3 - 84.4 (Finch, J. 72)
63 " (Mantelli) 120.4 + 73.5 ...nal, ...ß ...ßhaar ...
...n, glattschalig, ...ßig ... ß... (... Proc. J. 1. 59 - J. ... ...
62 " Owenii 105.7 + 63.3 ...ß, ...grau, ...
65 Charadr. bicincta 34.7 + 25.3 ...l. Algial. ...lt, ...
... ... ... ...
66 " falcus 48 + 33.4 48.1 + 33.5, ...n ...uratus (52 + 35.4)
(Seebohm - ...Tibir. /...) Ibis 79)
70 Thinornis (Haemat.) frontalis 36 + 25, ...
... ... ...
... ... ... ... ...
... 566 ...) - Harding, Proc. 74 - ...

...nalop. longirostris — 57.0 + 41.0 , buffy clone, mit großen innagalen. ...

... ... fast schwarzen ... ( Goals, B.a.)

　　 unicolor　　19.7 + 44.3 fahl ... farben innwall m. groß. innag...
　　 ... ..., ... ... liegen ... ... ( Goals, B...

...ea intermedia　42.5-52 + 33.4 - 38.4 - M: 48.1 + 36.4 ziemlich blaß ( Hume, ...
　 "　　　48.5 + 33.7 ... ..., ... blaß ... ( Campbell B.a.
　 "　　narra　40.2-47.7 + 31.5 - 33.6 - M. 43 + 32.9 ( Hume l.c.) ... blaß...
　　　　　47.5 + 31.7 blaß bläulich ... ( Goald l.c. )

...aurus pseciloptervs 50.7 + 37.9, ... ..., fahl ... ..., mit ... ...
　　 ... ( Campbell l.c. )

...icos. caledonicus　36.3 + 38　blaßgrün ( Goals l.c. )

... ... flavipes　71.3 + 45.9 ..., ... ... ..., ..., ..., ... ...
　　　64.5 + 45.2 , ..., ... ( coll. ... )

...antop. nov.zeel.　45 + 31 ... . ..., ... innagalen. ... ... ( ... ...
...ra uropygialis　56.9 + 36.6 blaß ... ... ... ..., mit ... ... ...
　　 ... ... ... ( Wales - Birds )

... ... ... 35.2 + 25.3 ... ... ... ..., mit ... ..., ...
　　 ... ... ... ... ..., ... ..., ... .. ..., ...
　　 ... ... ... ( Campb. )

...us ... Gr. (pectoral. ...) 37.9 + 30.3 ... ... o. ..., mit ... ...
　　 ... ... ( Hutton, Ibis 70.

...metra tahuenosis　29.5 + 23.2 , ..., ... ..., ... m. ...
　　 ... ... . ... ... . ( Campbell )

...nus australis　Größe ... ..., ... ..., mit ... ... ( Haast, H.b.
　　　54 + 39 , ... ... m., ( Nees. )

...hysio melanotus M. 50.7 + 37.9 ( Größe ... ...) fahl ... ..., ... mit ...
　　 ..., ... .. ..., .. ... ..., ... ... ... ... ... ...
　 "　　"　 53.2 + 37.9 ... ... o. ..., mit ... ... ...
...upervilova 59.5 + 44.2 ... ( Hutton )
　 "　　 59 + 40 ... ... , m. ... ... ( Layard, Ib. 82 )
　　 "　　 57 + 41 ... ... ( Goals )
...reps stectori 60.3 + 41.1 ... ... ... ... ( Campb. )
...nostyles patagonica 107.8 + 76, ..., ... . blaß ... ... ( Layard, Ib.
　　　103. + 79.5 ... ..., ... ( Thienem. )

57.0 + 41.0  buffy stone, mit großen Aurungalen. Latin schwarzen Blötik. (Gould, B.A.)

minutu, an 9.7 + 44.3  fall hurrfarben, überall u. groß. Aurungalen
The egg o  mer denen einige Linsen liegen u. zövgärf. sind (Gould, B.A.
Sula alba,
either spe  + 33.4 - 38.4 - M: 48.1 + 96.4  ziemlich blaß (Hume, N.s
           + 33.7  glfff.av., glatt, blaßfragrün (Campbell B.A.
           47.7 + 31.5 - 33.6 - M. 43 + 32.9 (Hume l.c.) sehr blaßfrag
    In N. 5 + 31.7  blaß bläulichmaiß (Gould l.c.)

represent  0.7 + 37.9, 3 gadn. eiförm., fall lank olivnaf., mit gaffer. Nam
to this bi
hawk; "tell l.c.)
bold little
comparat  6.3 + 38  blaßgrün (Gould l.c.)
chatterin  3 + 45.9  gefbn., ubf. 3. sp. f., maiß, nußschelig, fein gepart (Cam
the deniz
courage  it. 5 + 45.2, maiß, ängeff. (coll. Nehrk.)
penetrati  + 31 iful. s. Veaur., furfärung Aurungalen. Fröga ü. Al. (Hetting - n. Al.
ruthlessi
we have  9 + 36.6  blaß drab od. tief gmänl. drab, mit anwenaff. ul. pförpa
this dau  Automainen Zarfärungen (Water-Birds)
noticed
consider  35.2 + 25.3  fallfaurf. b. s. nafrigalb, mit olis. Auft., maiffl.
and kep  maffligen änkaurf. u. schmanz. Larveffl., Spr. u. Facsigala, dagmeiff.
    At  Camfab.)
breeding
some ex
of which  al. leir) 37.9 + 30.3  röthlichmaiß o. aufraf., mit Anfranimab
Pigeon.  (Hutton, Ibis 70.
    Se
examin  9.5 + 23.2  flaumfamal, schmätzig maiß, überall m. fall.
regards  l. u. Zigen. (Campbell)
sheltere
site for  einer Grfaarnit, gelblichmaiß, mit isokolm blaß Spr. (Haar, H.
peregrin  + 39, ein dinf. ook. m., (Nerk.)
deposit
collecte  7 + 37.9 (Großen f. nanismf. fall hurnänl. neffart., zänr. sich g
buildin  .9 +, m. Aurungal. mattflen., finern ü. Granzönig. Al. u. Spr.
mottle  (Campb
pale re  7.9  gelblichmaiß o. gelbbranän, mit röthl. Egnanst. A. u. fl. f.
broadly  (Hutton
1 inch  44.2  gelblichmaiß (Hutton)
Rakan  40  schmätzig nafrumaiß, m. gmänl. Auft. (Layard, H. 82)
place c
first, a  41  tinker nafufaur. (Gould)
spotte  41.1  gmänl. maiß ber gelblichmaänäa (Campb.)
Decem  8 + 76, birk lingly, schmätz. blaß gmänl. maiß (Layard, H.
the ey  + 79.5  gadn. ängeff., gelblich (Thienem.)
Potts,
    N
Rangi
    N
found
young
young

No. 2.—Circus Assimilis, Jard.

Kahu.

Harrier.

One of the commonest of the larger birds met with on "the plains." From its depredations on poultry of all kinds, game, etc., great numbers of this fine Harrier are annually destroyed by means of the gun, poison, or the trap. Over a lambing flock it may be frequently noticed soaring with wide circling flight. On a weakly lamb its attack commences by picking out the eyes. Birds it carefully plucks before it begins its meal. It is not an unusual occurrence to find it with a young flapper, almost as neatly plucked as though the work had been performed by the skilful hand of a poulterer. We found, on one occasion, a good sized Shag which had been thus operated upon; this was in winter time (July), and shows it has sufficient strength and courage to attack and destroy a bird of considerable size and power. Its favourite building-place appears to be a low-lying situation amongst swamps, the margins of lagoons, etc. The nest, built on the ground, is made of coarse grasses, such as tohe-tohe, raised sometimes about a foot in height, rather flat on the top. We have found it partly constructed with pieces of the thorny Discaria, and the dead flower-stems of the large Aciphylla, above which prickly materials grass has been carefully placed. The eggs, usually four in number, are white; when perforated, and held against the light, the interior shows a deep green; length, 1 inch 11 lines, with a breadth of 1 inch 6 lines.

A pair of these birds made use of the same nesting-place year after year, amongst some strong tohe-tohe, close to the Ashburton River. We took from this nest an egg, which had been entirely covered up with the materials which had been brought to renovate the nest, at a period, subsequent to the breeding time, at which this egg had been laid.

From our memoranda, the months of November and December appear to be the height of the breeding season; it is found moulting in February; occasionally fine old specimens are met with, in which the whole plumage has assumed quite a light tone of colour; this is so conspicuous in some individuals, that some collectors endeavour to persuade themselves that a new species has been discovered. Perhaps the noiseless flight of this bird should be noted. When swooping on its quarry, the clean long tarsi enable the observer to see the action of the feet, the rapid contraction and expansion of the toes, when striking at its prey; should this prove too large, or too heavy, to be swooped off at once, the Harrier will drag it a considerable distance, apparently changing its hold frequently, accompanied with much noiseless fluttering of the wings, each time it strikes out its sharply armed foot to obtain a fresh grasp. To give some idea of the numbers of this hawk that are annually destroyed, it may be mentioned, that on the Cheviot Hills station, ten to twelve per day were frequently killed, and that it would be within compass to reckon that upwards of 1,000 hawks per annum had been thus accounted for during the last two or three years; amongst these were a few of the *Falco N.Z.* It will not create surprise to learn, that on this run rats are most abundant. On a farm on the Halswell, as many as fifteen were found poisoned in one morning. On another farm in this neighbourhood, numbers have been trapped by the use of a common rat-gin fixed on the top of a Ti palm.

No. 7.—Halcyon vagans, Gray.

Kotare.

Kingfisher.

One of our burrowing species. The tunnel-like hole, which forms the approach to its nest, is found sometimes in a bank, and, perhaps, quite as often in a tree. On examining one of these holes, in a bank not far from the sea beach, the floor or bottom was observed to incline slightly upwards from the entrance, the eggs, deposited on the remains of crustaceæ, were not more than one foot back from the outside of the hole. When a tree has been selected for its home, we have been led sometimes to the discovery, by observing the quantity of chips lying beneath ; its powerful bill soon excavates a nesting-place in the partially decayed wood. The situation varies from a few feet to above thirty feet from the ground (See Plate 4, Fig. 1). The eggs are pure glossy-white, delicate, and very beautiful, more fragile, perhaps, than those of most other species, oval in shape, with a length of 1 inch ½ line, by a breadth of 10½ lines. After hatching, the nest is carefully cleared of the broken shells. The young remain in the nest till well-fledged, and, apparently, almost full-grown. On examining the castings of the Kingfisher, which are often to be met with in abundance near a nest containing young, we have observed that the external wing-cases of coleopteræ, have formed one of the principal ingredients of the pellets. We have noted that a nest from which the young emerged late in November, again contained eggs in January. Our Halcyon must lay a much smaller number of eggs than the English Kingfisher. Although this bird may be constantly seen occupying some prominent branch, or stake, when watching for its prey (which, by the way, is of a very miscellaneous character), yet, when approaching or leaving its nest, it always, where possible, seeks the screen of overhanging trees, as it swiftly darts through the gully, permitting but a glimpse of its bright showy feathers. Should any one approach too close to the neighbourhood of its breeding-hole, the parent bird utters a low cry, like cree, cree, cree, frequently repeated. Our bird is much more sociable than its European relative, which is so remarkable for its solitary habits, that it has been stated, that the male and female only associate together at the breeding season : we have counted as many as eight of our Kingfishers sitting in company ; after a heavy rain we have observed, on our lawn, several of the croquet hoops occupied at one time by these striking-looking birds. It is rarely to be seen on the ground ; after darting down, either in the water, or on land, and securing its booty, it immediately flies with it to some perch, or post of vantage, and prepares it for deglutition, by administering some smart blows with its bill, the sound of which may often be distinctly heard. During the breeding season it indulges in a monotonous call of chimp, chimp, chimp, then a pause, the call and pause alternating for a considerable time. Fish, crustaceæ, young birds, mice, coleopteræ, bees, and other insects, furnish some portion of the food-supply of the Kingfisher ; we have often noticed its rapid dart at a brood of young chickens. This bird is one of those fortunate species, whose numbers seem rather to increase than diminish at the approach of civilization.

The name of Halcyon given by ornithologists to this species, carries us far back into the very early days of Natural History. The history of its European congener was enveloped in poetic fables for centuries ; probably no other bird, whose habits could be so easily observed, has been so universally the subject of groundless tales, or superstitious regard,—perhaps the recital of some of these notices may be excused. Aristotle, after a fair description of the bird, gravely states : " Its nest resembles the marine balls which are called *halosachuar*, except in colour, for they are red ; in form it resembles those *sicyæ* (cucurbits) which have long necks." Again, he says : " This bird hatches its young about the time of the winter solstice. Whereupon fine days occurring at this season are called *Halcyon* days." Omitting the fabulous accounts of many ancient authors, let us peruse the account of the philosopher of a more recent date, on the breeding habits of this wonderful bird ; thus quaintly wrote Montaigne :—

" Mais ce que l'experience apprend à ceux qui voyagent par mer et notamment en la mer de Sicile, de la condition des halcyons, surpasse toute humaine cogitation. De quelle espèce d'animaux a jamais Nature tant honoré les couches, la naissance, et l'enfantement ? car les Poëtes disent bien qu'une seule isle de Delos, estant auparavant vagante, fut affermie, pour le service de l'enfantement de Latone : mais Dieu a voulu que toute la mer fut arrestée, affermie, et applanie, sans vagues, sans vents, et sans pluye, cependant que l'halcyon fait ses petits, qui est justement environ le Solstice, le plus court jour de l'an : et par son privilege nous avons sept jours et sept nuicts, au fin cœur de l'hyver que nous pouvons naviguer sans danger. Leur femelles ne recognoissent autre masle que le leur propre : l'assistant toute leur vie sans jamais l'abandonner : s'il vient à estre debile et cassé, elles le chargent sur leurs espaules, le portent partout, et le servent jusques à la mort.

" Mais aucune suffisance n'a encore peu atteindre à la cognoissance de cette merveilleuse fabrique, dequoy l'halcyon compose le nid pour ses petits, ny en deviner la matiere. Plutarque, qui en a veu et manié plusieurs, pense que ce soit des arestes de quelque poisson qu'elle conjoinct et lie ensemble, les entrelassent les unes de long les autres de travers, et adjoustant des courbes et des arrondissemens, tellement qu'enfin elle en forme un vaisseau rond prest à voguer : puis quand elle a parachevé de le construire, elle le porte au batement du flot marin, là où la mer le battant tout doucement, luy enseigne à redouber ce qui n'est pas bien lié, et à mieux fortifier aux endroits où elle void que sa structure se desment, et se lasche pour les coups de mer ; et au contraire ce qui est bien joinct, le batement de la mer le vous estreinct, et vous le serre de sorte, qu'il ne se peut ny rompre ny dissoudre, ou endommager à coups de pierre, ny de fer, si ce n'est à toute peine. Et ce qui plus est à admirer, c'est la proportion et figure de la concavité du dedans : car elle est composée et proportionée de maniere qu'elle ne peut recevoir ny admettre autre chose, que l'oiseau qui l'a bastie : car à toute autre chose, elle est impenetrable, close et fermée, tellement qu'il ny peut rien entrer, non pas l'eau de la mer seulement. Voyla une description bien claire de ce bastiment et empruntée de bon lieu : toutesfois il me semble qu'elle ne nous esclaircit pas encor suffisamment la difficulté de cette architecture. Or de quelle vanité nous peut il partir, de loger au dessous de nous, et d'interpreter desdaigneusement les effects que nous ne pouvons imiter ny comprendre ?"

Sir Thomas Browne, the exposer of vulgar errors, in his " Pseudodoxia Epidemica," after stating the results of actual experiments, which enabled him to contradict the common notion, that a Kingfisher, suspended by the bill, would show from what quarter the wind blew, yet, apparently, received the ancient fable of the halcyon days without any distrust, for thus he wrote concerning the peculiar relations existing between this bird and the winds :—
" More especially remarkable in the time of their nidulation and bringing forth their young. For at that time, which happeneth about the brumal solstice, it hath been observed, even unto a proverb, that the sea is calm, and the winds do cease, till the young ones are excluded, and forsake their nest, which floateth upon the sea, and by the roughness of the winds, might otherwise be overwhelmed. But how far hereby to magnify their prediction we have no certain rule ; for whether out of any particular pre-notion they choose to sit at this time, or whether it be thus contrived by concurrence of causes, and providence of nature, securing every species in their production, is not yet determined." It would occupy too much space to mention the names of naturalists and writers who adopted similar romantic tales, each of whom was, of course, supposed to be narrating a particular and veracious account of the extraordinary mode of nidification of the Halcyon. Mr. Gould dissipated,

at last, whatever might have remained of these clouds of fable, by depositing the nest, entire, in the British Museum; a feat, the difficulties attending which were so well appreciated by all bird-nesters, that there was a report, or tradition, throughout many parts of England, that the authorities of the British Museum had offered a reward of £100 for a perfect nest of the Kingfisher. For a full account of Mr. Gould's exploit, see " Homes without Hands."

Shakespeare, in " King Lear," and several other writers, allude to the superstitious idea, that, if suspended by a thread from the ceiling, with windows and doors closed, the Kingfisher would turn its bill towards the quarter from whence the wind blew.

Amongst numerous other virtues, it was supposed to be a protection against thunder, against the ravages of the moth in woollen cloth, to be able to increase hidden treasure, to bestow grace and beauty on the person who carried it, and enjoyed the power of renewing its plumage, after death, by moulting.

## No. 10.—Prosthemadera Novæ Zelandiæ, Gml.
### Tui.
### Parson-bird.

We have but seldom found the nest of this very common bird, whose varied notes break upon the stillness of the bush. Wherever we have met with its nest, it has been rather on the outskirts than in the depth of the bush itself. The Parson-bird seems thoroughly joyous only in the full glow of sunlight, where it may be seen in numbers, darting upwards far above the highest trees, and revelling in its free stretch of wing, now and then playfully pursuing some smaller bird, till it seeks the shelter of a friendly bush.

We have found the nest from twelve to thirty feet from the ground, and have noticed that whether against a White pine, or Black birch, there has been a sheltering cluster of Rubus, with its sharp, recurved prickles, beneath which the structure has been concealed. We have found it more than once near the top of a *Myrsine Urvillei*, over which the Rubus has thrown its straggling cords, forming a prickly canopy most difficult to penetrate. The nest, rather large, made of slender sprays intermixed with moss, and the wool or down of Tree-ferns (*Cyathea dealbata*), lined with fine bents of Poa grass; the dimensions we noted of a nest are as follows: across the top, from outside of wall to outside of wall, 9 inches, diameter of cavity, 3 inches 6 lines, with a depth of 2 inches. The eggs, usually three or four in number, are white, or with the slightest tinge of pink, marbled with rust-red veins, most numerous towards the larger end, rather pyriform in shape. they measure 1 inch 2 lines in length, by 10 lines in breadth. The nest containing young is sometimes stained deep purple, from the juice of the Konini berries (*Fuchsia excorticata*). On one occasion, the young, unable to fly, on being alarmed fluttered out of the nest to the ground, a fall of about twelve feet, the next day they were found safely ensconced within the nest, looking quite happy; this could only have been effected through the assistance of the parent birds. The Tui is rather combative whilst the young require feeding, even when they can fly well, it may be observed driving away the Kingfisher and Bell-bird from the trees in which its young are lodged. However much the white-tufted Tui may add to the interest of our forest scenery by the beauty of its glossy plumage, the gaiety which distinguishes its flight, or the wild outburst of its joyful notes, in the eyes of the omnivorous settler, it possesses the higher merit of furnishing a savoury article of food, and no weak sentimental feeling saves it from the camp-oven. It is frequently kept in confinement. and at one time many were sent to the neighbouring colonies. (See Plate 6, Fig. 1).

Fig. 3.

Fig. 4.

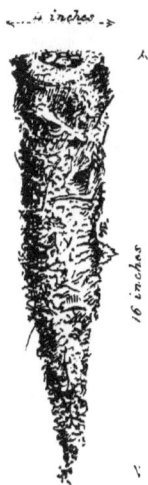

Nests of
## PETROICA MACROCEPHELA.

Fig. 7.

Nest of
## BOTAURUS POICILOPTERUS.

Bibburn.

Fig. 1.

Nest of
HALCYON VAGANS.
King Fisher.

Fig. 2.

Nest of
ACANTHISITTA CHLORIS.
Wren.
Built in a small roll of bark hanging in a cluster of Convolvulus.

Fig. 3.

Fig. 4.

Nests of
PETROICA MACROCEPHELA.

Fig. 5.

Nest of
PODICEPS HECTORI.
Grebe.

Fig. 6.

Nest of
RHIPIDURA FLABELLIFERA.
On a frond of the silver tree fern  Cyathea dealbata.

Fig. 7.

Nest of
BOTAURUS POICILOPTERUS.
Bittern.

## No. 11.—Anthornis melanura, Sparm.

### Koromako.

### Bell-bird.

Everyone who has rambled through the bush, or even strayed amongst the shrubby thickets that fringe our numerous gullies, must have become familiar with the clear metallic ring of the Bell-bird's note. It may be said to sing matins and vespers for the warblers of the bush, as it is at the grey break of dawn, and the still hour that closes in the day, that its chime strikes clearest on the ear. It is comparatively silent during the noontide heat, unless some few individuals meet on a tree or shrub, that offers a tempting show of honey-bearing blossoms, a note or two is briefly sounded, the numbers rapidly increase; after much noisy fluttering of wings, a gush of clanging melody bursts forth from a score of quivering throats, forming a concert of inharmonious, yet most pleasing sounds. Probably Cook indicated the Bell-bird, then in a comparatively unmolested state, when he wrote, "the ship lay at the distance of somewhat less than a quarter of a mile from the shore, and in the morning we were awakened by the singing of the birds; the number was incredible, and they seemed to strain their throats in emulation of each other. This wild melody was infinitely superior to any that we had ever heard of the same kind; it seemed to be like small bells, most exquisitely tuned, and perhaps the distance and the water between, might be of no small advantage to the sound." Nor does this cheerful bird confine itself to the bush, it frequents our gardens and shrubberies, and especially affects the blossoms of the Fuchsia, Tritoma, Acacia, etc. The berries of various Coprosmas, and that of the Konini, it greedily devours; it may be frequently observed fluttering heavily in pursuit of a moth. It is very easily snared with a noose at the end of a tohe reed; in confinement it feeds on soaked bread, etc. Whilst the *Phormium tenax* is in blossom, many Bell-birds may be observed with their head feathers dyed orange-red, from contact with the pollen and honey, whilst extracting a delicious repast from the flax blooms. It has been stated that zealous ornithologists have deemed the bird thus decorated, a new species.

Placed at no great elevation from the ground, the nest may be found in a variety of positions, but we certainly have noticed it most frequently beneath a sheltering canopy of "Bush-lawyer" (*Rubus australis.*) It is rather flat, and loosely constructed of sprays, grass, moss, etc., well lined with feathers. On examining the foundation of a nest, we found green sprays of Manuka amongst the interlaced materials, a fact which disclosed the proof of the power of the bill of this honey-sucker in breaking off such tough twigs. From wall to wall, across the top, the nest measures about 5 inches, diameter of cavity, 2 inches 9 lines, depth inside, about 2 inches. We fancy that the lining feathers are selected in such a manner as to afford some evidence of harmony of colour in their arrangement; as, for instance, we have noted specimens, with the inner lining entirely composed of the red feathers of the Kaka, another adorned with the green feathers of the Parroquet: near the farm, where many kinds of poultry are kept, we have had instances of lining, white, black, speckled, buff, etc., but uniformity of colour has been displayed. The eggs, four in number, are white with reddish specks, sometimes the ground-colour exhibits a delicate pinkish tinge; they measure in length 11 lines, with a breadth of 8½ lines. We must have peered into scores of nests, in various parts of the country, but we have never yet been fortunate enough to encounter such a prize as one containing "seven eggs, spotted with blue, upon a brown ground," ascribed to this bird by the Rev. R. Taylor, in his work "Te Ika a Maui." The breeding-season extends from the commencement of spring, throughout the summer

c

months. We have discovered the nest in an old flower-branch of the Ti palm (*Cordyline australis*). (See Plate 5, Fig. 1.)

Note.—Feb. 2, 1868—Bell-bird building; that would give the breeding season a duration of about six months.

## No. 15.—Pogonornis cincta, Dubus.

A nest, assigned to this bird, was found in the bush above the Kaiwarawara stream, not far from Wellington; it contained one egg, rather oval in form, somewhat pointed at each end, measuring 9 lines in length, with a breadth of 7 lines; the whole surface clouded over with pale rufous-brown.

The nest, with thin walls, and of shallow form, was built of sprays, above which were laid fibres and dry rootlets of Tree-fern; fine grass was used for the lining, over which cow-hair was laid, and measured, across the top, 4 inches 9 lines, cavity 2 inches 4 lines, depth 1 inch 4 lines. This description is from the specimen in the Colonial Museum, Wellington.

## No. 18.—Acanthisitta chloris, Sparrm.
### Pi wau wau.
### Wren.

This, the smallest of our birds, is usually seen in pairs, flying low, with a feeble, jerky style of flight; more frequently it is met with creeping amongst the lichens and mosses that decorate the stems and branches of our forest trees. We have found the nest in a small hole in the trunk of a Fagus. Once a nest was discovered, very cleverly built in a roll of bark, that hung suspended in a thicket of climbing Convolvulus. (See Plate 4, Fig. 2).

The eggs are said to be very numerous sometimes, although four or five have been the most we have observed to a nest; like those of nearly all troglodytal birds, they are white and glossy; ovoiconically shaped, they measure $7\frac{1}{4}$ lines in length, by 6 lines broad. We have a note of the Wren breeding in August.

## No. 19.—Mohoua ochrocephala, Gml.
### Mohoua.
### Canary.

Although we have not observed this bird anywhere on "the plains," or on the lower ground of the "bays," yet as soon as one ascends the bushy gullies of the hills, the Canary is sure to pay a reconnoitering visit; with sharp strident call, it summons its companions, and the trees around will soon disclose the golden breasts and heads of these active arboreals, as they peer down on the intruder with noisy clamour. With restless movements, they creep round, above, and below the leafy branches, in their insect search. We have watched them on the ground, busily scratching and pecking between the huge moss-clothed roots of the lofty trees that tower above. The nest measuring across the top, 3 inches 3 lines, with a depth of 1 inch 4 lines, is a beautifully compact structure, cup-shaped, principally of moss, very closely felted, and neatly interwoven with webs of spiders. (See Plate 5, Fig. 2). In the hollow trunk of the Broad-leaf, it is sometimes found, and occasionally in a decaying Black Birch. Eggs white, with very small faint specks of red, nearly 11 lines in length, with a breadth of $8\frac{1}{2}$ lines. We have a specimen of the nest and eggs from the River Wilberforce.

## No. 20.—Sphenœacus punctatus, Quoy. and Gaim.
### Mata.
### Grass-bird, Grass-pheasant, Utick.

Some years ago the monotonous note of this little bird might be heard in almost any place, where tall tohe-tohe reeds, or the waving leaves of the *Carex virgata*, indicated marshy ground ; now it is rapidly disappearing, as the swamps are becoming drained. As its very feeble power of flight is unable to save it from the bush fires, we anticipate it must become extinct, on " the plains," at no very distant date. From its call, it is in some places named the Utick. The nest, inclining somewhat to an oval shape, and measuring about three inches across, is made of grass leaves, so frail in its construction, that the walls may be seen through, consequently it is a difficult specimen to obtain in a perfect state (See Plate 5, Fig. 4) ; a few feathers, usually those of the Pukeko, are added to the grass leaves, and sometimes a small tuft or two of wool. The situation is, most frequently, in a tussock, a few inches above the level of the ground. The eggs, three or four in number, are white, speckled with a beautiful tint of reddish-purple, which at once readily distinguishes them from those of any other bird ; ovoiconical in form, they measure, through the axis, 10 lines, with a diameter of $7\frac{3}{4}$ lines.

Notes.—Nov. 4—Nest containing three young birds, in a tussock, at the edge of a wide creek.

Nov. 7—Nest with four eggs, in a swamp by the Hororata stream, in the Malvern Hills.

### No. 25.—Gerygone assimilis, Buller.
### Piripiri.
### Warbler, Teetotum.

This cheerful little warbler is a pensile nest-builder, and one of the earliest breeders ; its neat, domed nest may be often found, in August, suspended in some bushy Manaka or Olearia. The nest may be called somewhat pear-shaped, with a small entrance near the middle, above which is often affixed a kind of porch (See Plate 6, Fig. 3), it is suspended by its top, and kept steady from swaying in the breeze, by slight fastenings to a spray or two, acting as guys. Moss enters largely into its construction, very frequently wool ; we have examined one, the greater part of which was composed of wool ; cobwebs are freely made use of, to felt and bind the materials into a compact mass. We have a nest before us, taken from the fork of a Willow tree, at least twenty-five feet from the ground ; it is rather larger than usual, and almost wholly constructed of poultry feathers and cobwebs, and is felted into a compact, firm structure, the porch and its foundation, beneath the entrance, is strengthened and kept in shape by fine roots carefully interwoven with green cobwebs ; here and there may be found pieces of thread, string, coloured worsted, picked up from the garden or yard ; the interior is thickly lined with feathers (See Plate 6, Fig. 2), this nest is evidently composed of materials, which would not have been made use of so freely, but for its firm and sheltered position in the fork of the willow, the most exposed part only being strengthened with stiff material.

Sometimes, yet rarely, the nest is built in a less elaborate manner, without either dome or porch, the form of the structure being adapted to the peculiarities of the situation chosen ; the principle of suspension is likewise occasionally abandoned. Five or six eggs are usually found to a nest, they are white, with red spots, ovoiconical in shape, 8 lines in length, with a breadth of 6 lines. No bird suffers so frequently from the imposition of the golden-winged Cuckoo, as the grey Warbler. We have several times observed a pair of these industrious little insect-eaters, feeding a young parasite larger than themselves. The Cuckoo only arrives in October, when the warmth of Spring is well established ; and one reason for the selection of the Warbler's home, in addition to its pencile character, appears to us to be, because from its shape and structure it is the warmest nest, to be found, for rearing so tender a bird as the Chrysococcyx, our gay visitor, during the spring and summer months.

Note.--We have found eggs of the Warbler quite white, doubtless the produce of young birds. As yet we have failed to observe any such distinctive features, either in the structure or habits of these Warblers, that they should be classed as separate species, under the names of *flaviventris* and *assimilis*. We adhere to *assimilis*, as is adopted in the collection in the Canterbury Museum.

## No. 26.—Certhiparus Novæ Zelandiæ, Gml.
### Brown Creeper; Brown Canary.

Although this Creeper may be seen in almost every bush, from the coast to the distant Alpine Ranges, we have only once found its nest. This was in the month of December, far above the Rangitata Gorge. The nest, containing three young birds, was compactly built of moss, with a few feathers, placed in a Black-birch, between the trunk and a spur, from whence sprouted out a thick tuft of dwarfish sprays, about seven feet from the ground.

## No. 27.—Certhiparus albicilla, Less.
### Mohoua.

This bird appears sufficiently common, about the bush above Wellington, for its habits to be well studied. There are several specimens of the nest and eggs in the Colonial Museum, Wellington. The nest is a very compact structure, having very thick walls, and in its style of architecture bears a strong resemblance to that of *M. Ochrocephala*, although, in some instances, different materials are used. In the one before us, different kinds of soft grass and moss form the staple, well-felted and interwoven with webs, lichens, and the down of tree-ferns; it measures 4 inches 1 line across the top, cavity 1 inch 10 lines in diameter, 1 inch 4 lines deep. Eggs white, or with very faint specks of pink, measure 10½ lines in length, with a breadth of 7½ lines.

## No. 29.—Petroica macrocephala, Gml.
### Ngirungiru. Piro piro.
### Tomtit.

This familiar little bird is one of the more elaborate nest-builders amongst the denizens of the bush, or rather its outskirts.

It adapts itself, in a manner, to civilization, frequenting gardens, and may be seen perched on a bough, ready to pounce on the grubs the gardener's spade may bring to light, reminding one very much of the habits of the Red-breast at home.

The nest varies much in shape according to position; frequently we have found it in holes of trees; a favourite site is immediately under the head of the ti tree (*Cordyline australis*). Two nests we presented to the Canterbury Museum, were of remarkable shape; one, a firm compact structure, placed in the forked head of a ti tree, resembled a very neat moss basket, with a handle across the top; the second, also from a ti tree, from, perhaps, the foundation slipping between the leaves, was built up till it reached the great length, of sixteen inches. (See Plate 4, Fig. 4). We have found others placed on a rock, and one, now in the Colonial Museum, was built between the brace and shingles in the roof of an empty cottage.

The nest is neatly and firmly built of a variety of materials, carefully and neatly interwoven; moss, grass-bents, slender sprays, the down or wool of the tree-fern, cobwebs, and feathers, warmly line the interior. Four eggs is the usual number laid, though we have been told of more having been found; they are white, with grey speckles, most numerous towards the larger end, 9 lines long and 7 lines broad. A nest built in a ti tree, close to a pathway, was almost daily visited by the child who had made the discovery, and the eggs inspected ; when hatched the young were now and then handled, yet the confidence of the old birds carried them through this trying ordeal, and their young ones were successfully reared.

This is one of the few birds, of whose extinction we are happy to believe there is no danger ; it is most useful as an insect eater, it is one of the latest to retire to rest, and may be often observed perched on the trunk of a tree, in a posture by which its body is almost at a right angle with the tree. The nests, described above, were found about Ohinitahi, where birds are as much encouraged, and as little disturbed as possible. Last summer another specimen was noticed, which had been built upon an old nest, making a solid mossy structure, measuring about one foot from top to bottom. The usual dimensions of the nest are as follows :—Across from outside of wall to outside, 5 inches ; cavity 2 inches 6 lines, with a depth of 1 inch 6 lines.

### No. 31.—PETROICA TOITOI, Less and Garn.
### Tit.

Whatever distinguishing features, scientific research may have discovered, which allows specific differences between *P. Dieffenbachi* and *P. toitoi*, we fear they are not generally appreciated or understood. Perhaps this may be a fair opportunity of pointing out that the nomenclature of our birds still requires attention, and, above all, *settlement ;* to the enquiring student of ornithology, scarcely anything can exceed the perplexity and embarrassment which is caused by a conflicting nomenclature. To give one instance : *Anarynchus frontalis* appears in Dieffenbach's list ; since then we have noticed it as *Charadrius, Hæmatopus,* and now *Anarynchus* once more. Let us hope this may be the last change. We have often observed a Petroica, whose favourite haunt appeared to be amongst large areas of flax bushes (*Phormium tenax*), but confess we could not undertake to decide to which of the two species, named above, the Tit, to which we have referred, belonged ; nor is there, unfortunately, any complete type collection, either in Wellington or Christchurch, which could decide any doubt that might be entertained on the subject. We have a set of eggs in our collection, which we are inclined to assign to the *P. toitoi ;* they are slightly more inclined to pyriform, in shape, than those of *P. Macrocephala,* white, with marks of purplish-grey towards the larger end, and measure 9 lines in length, with a breadth of 7 lines.

### No. 32.—PETROICA LONGIPES, Less. and Garn.
### Robin.

In the Colonial Museum, Wellington, there is a specimen of the nest and eggs of this bird.

The nest, compactly built of moss, fine roots, web, and tree-fern down, is more neatly finished than that of *P. albifrons.* The eggs, ovoiconical in form, are marked, principally at the larger end, with specks of greyish-brown.

### No. 33.—PETROICA ALBIFRONS, Gml.
### Totoara.
### Robin.

Our rather dirty-looking Robin is one of the sweetest warblers of the bush, bold and confident, its habits may be easily observed, as one rambles near the rocky sides of a forest stream. Its nest is wider, and larger altogether, than that of *Petroica macrocephala,* but not so closely interwoven ; moss, sprays, leaves, fine fibres, and grass, enter into its construction. Diameter of nest 5 to 6 inches, cavity 3 inches, with a depth of 1 inch 3 lines. A favourite situation appears to be behind such protuberances as are to be found on the huge gnarled trunk of *Griselinia litoralis*, very often not more than three feet from the ground. Eggs, three or four in number, are dullish-white, with reddish marks, principally at the larger end.

## No. 34.—Anthus Novæ Zelandiæ, Gml.
### Pihoihoi.
### Lark.

This well-known bird appears to be common all over the country; it builds on the ground, making its nest of grass, usually screened by a tussock. The eggs, five in number, are greyish-white, speckled over with dark-grey; sometimes a set of eggs may be noticed very much mottled over with brown, ovoiconical in form, measuring 10½ lines in length, by a breadth of 8 lines. We have an egg, very much smaller and darker than any others we have yet observed. In February, 1868, a pair made their nest within six inches of a shrubbery walk, and reared their young successfully, although so frequently disturbed,—the old bird invariably quitted the nest on its being approached. When a Harrier wheels round, and appears about to settle, Larks may often be observed, in numbers, gathering together with a chirping note, moving restlessly, sometimes with a short flight, watching and following the movements of their enemy.

Probably it is attempting to rid itself from the persecution of some parasitic vermin, when this bird is frequently observed to indulge in a dust-bath. It has a habit of keeping its insect prey in its beak for a long time, before it is devoured, or carried off to its nest. At last shearing time, two Larks, almost albinos, made their appearance, daily, about the yards of a wool-shed, on the Waikerukini.

Note.—In August, a nest was brought to the Wellington Museum, which contained several tufts of moss, but not neatly interwoven, like the workmanship of a bird that builds its nest principally of moss.

## No. 35.—Zosterops lateralis, Lath.
### Tauhou.
### Blight-bird.

We first noticed this bird on some Fagus trees in the Rockwood Valley, Malvern Hills, July 28th, 1856. Its numbers, since then, have increased with great rapidity. It very soon obtained the name of the Blight-bird, in recognition of its services to gardens and orchards, from its habit of feeding on the American blight, with which apple trees in this colony are so generally infested; but, although the gardener noticed with satisfaction its labours in this direction, during the winter months, yet as summer returned and fruits ripened, its incessant depredations on cherries and plums were witnessed with anything but pleasure. From examining scores of nests, we find that out of a considerable variety of materials made use of, moss and grass predominate; the fabric is strong. although frequently slight, in some cases the walls are extremely thin; it is usually suspended, at the sides, by fastenings bound securely over slender twigs; some are almost wholly constructed of grass, amongst which, now and then, may be found a few small tufts of the grey-beard moss, in others the cottony down of plants is neatly interwoven with moss and spiders' webs, lined with fibres, or fine stems of grass, sometimes with hair; some nests are quite shallow, others of deep cup-like form (See Plate 5, Fig. 3), and measure in diameter 3 inches, cavity 1 inch 6 lines to 2 inches, depth 10 lines to 2 inches. In gardens, it has been observed placed in a great variety of shrubs, occasionally in a rose-bush bordering a well-frequented walk;

never far above the ground, usually from two to six feet. We have found it suspended to our common fern, *Pteris aquilina*. It lays three clear-blue eggs, ovoiconical in shape, measuring 8 lines in length, with a breadth of 6¼ lines: incubation lasts about ten days. The nest and eggs form as pleasing an object as those of the Hedge-sparrow at home. The gift of song does not appear to be equally shared by these birds; in addition to the quick, sharp note or chirrup, which all seem to have in common, now and then an individual bird is heard pouring forth a low, well-sustained, melodious song; possibly the power may exist in all adult males, only to be indulged in at pairing time.

One of the pensile nest-builders, which seem to be almost equally rare in our temperate clime as they are in the old country. The suspension of its habitation is accomplished in a different manner from that of Gerygone, and more after the fashion adopted by *Regulus cristatus*, of Europe, the Kinglet or Golden-crested Wren, except that the nest is very often formed without any protection or shelter from an overhanging leaf. The rim of the ladle-shaped structure is firmly secured to a forked twig by silky threads of spiders' nests, finished on the outside, round the bottom, with braces of green leaves of grass, crossed and recrossed, which add much to the strength and stiffness of the fabric.

Now, as pensile nests are peculiarly adapted for ensuring the safety of their contents against the predatory attacks of various egg-robbers, does not the suspension of the habitation of the Zosterops,—the instinctive precaution of a foreign land (See *Chrysococcyx lucidus*).—afford an indication that it is a recent colonist, not yet so thoroughly acclimatized as to be fully aware of the immunity it enjoys from ravages of snakes, etc. ? will that form of nest which is now sometimes found built *in*, rather than suspended *from*, a bush or thicket, become a more common object, and thus show a change in the style of architecture, as this bird, season after season, experiences the comparative safety of the breeding-places in our cooler latitude ? Amongst our indigenous genera are there any pensile nest-builders ? For years we invariably found three eggs to be the complement to a nest; now this last season we have met with several instances where four eggs have been laid, where this has occurred, the home has been built *in*, rather than fairly suspended *from*, a bush. If the reason, before suggested, for a modification in the manner of fixing the habitation be considered as not altogether too fanciful, may we not likewise be allowed to advance our opinion that the change of climate is also gradually producing its effects in the increased fecundity of our little Blight-bird.

Note.—Dec. 4—Nest in a manuka (*Leptospermum scoparium*), appeared to be completely lined and finished. On the 8th it contained three eggs; the next day a fourth egg was laid; on the 19th one callow nestling was exhibiting its ugliness, perfectly naked, except two or three small tufts of white down on the bald cranium, the body deep yellowish-pink, with dark slatey-coloured marks along the line of the vertebræ, the exterior of wing, and legs. The day following, his ugliness had a companion, on the 23rd feathers had made their appearance, where the slatey markings had been noticed; two unhatched eggs remained in the nest, which was only visited quietly once a day.

Young birds, for some time after they can fly well, can scarcely be said to possess any just pretension to the title of Zosterops, as they are without the circlet of white feathers round the eyes.

From the large number of nests we have observed, December must be the height of the breeding season.

The Zosterops is so partial to the berries of the trailing *Cotoneaster mycrophylla*, that we have known it to be taken by the hand, when it has been busily engaged on them; in the early spring we have observed it eating clover.

## No. 37.—RHIPIDURA FLABELLIFERA, Gml.
### Piwakawaka.
### Fan-tail.

The pied Flycatcher seems to prefer proximity to water in selecting its nesting-place, we have noticed it most frequently near a creek, where over-hanging boughs have afforded considerable shade.

The nest, beautifully made, is very compact, and, from our experience, varies very slightly in shape. The materials are well felted together, moss, grass-bents, fibrous roots, with cobwebs, etc. ; the structure is fixed on some bough or spray, the foundation, very frequently, commences with chips of decayed wood. The prettiest nest we ever found, was on a leaf of the large silver tree-fern (*C. dealbata.*) (See Plate 4, Fig. 6.) The eggs, four in number, generally are white with brown freckles towards the larger end, 8 lines long, by 6 lines broad. We never found the nest very early in the spring.

Towards autumn this bird frequents the verandah, enters the house, clearing the rooms of flies, the snapping of the mandibles is plainly heard, as it flits circling round the room.

*R. albiscapa*, the fan-tail warbler of Tasmania, builds a nest with a long tail underneath, giving the whole structure a funnel-like appearance. Occasionally, *R. flabellifera* also builds its home with a long tail, but broader and less artistically finished than that of the *R. albiscapa*. One nest in our collection has this peculiar appendage, constructed of skeleton leaves and bents of grass, etc. What is its use ?

### No. 38.—RHIPIDURA FULIGINOSA, Sparrm.
### Tiwaikawaka.
### Black Fan-tail.

The Black Fan-tail Flycatcher breeds under conditions so very similar to those of the preceding species, that one description will serve for both. To our view, the most remarkable feature in the breeding habits of our Flycatchers is the situation usually selected for rearing their young. Security does not appear to be the first consideration ; security, by concealment, seems the leading feature which guides most arboreal birds in choosing the site for their home, and it is one in which the most admirable displays of instinct may be frequently observed. The Flycatchers rather appear to be led by the same consideration which actuate many sea-birds in selecting the position of their breeding-place, proximity to the food supply. Stroll carefully along the rocky bed of a creek which rambles through some bushy gully, and you may perchance see the beautiful nest perched on some slender bough, in so delicate a manner, that it appears scarcely so much to be fixed, as to rest balanced there. There is no concealment amongst tangled creepers, guarded with their sharp recurved prickles ; it is not buried amidst a mass of waving leaves, nor is it hidden away in the dim twilight of some hollow tree, but there, a few feet above the water, it sways gently with the subdued breeze, that reaches the quiet ravine through the leafy canopy that is spread around.

In thus placing its nest so obviously in view, one is reminded of its family connections, of the Spotted Flycatcher (*Muscicapa grisola*) of the old country, which we used to term the Post bird, from the almost glaring manner in which its unscreened habitation was displayed. But as " there is reason in the roasting of eggs," saith the proverb, so there is also instinct in selecting the place where they shall be laid ; over the shady creek our Flycatcher is in the midst of sandflies, and the position chosen for its nest affords comparatively as good a vantage ground for supplying the wants of its young, as the nesting-place on the craggy mountain side bestows on the dashing Quail-hawk.

Fi

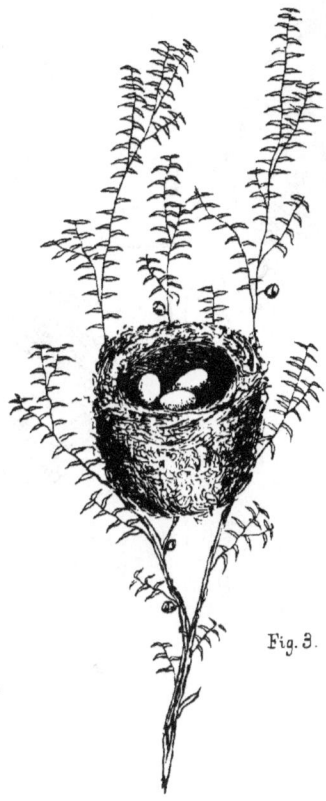

Fig. 3.

Nest and Eggs of

ZOSTEROPS LATERALIS.

*In a branch of Leptospermum scoparium.*

To accompany Paper
by
T. H. POTTS.

*In a*

Fig. 1

Nest and Eggs of
ANTHORNIS MELANURA.
*In an old flower spike of Cordyline Australis.*

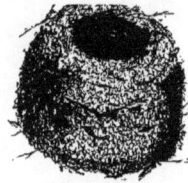

Nest of
MOHOUA OCHROCEPHALA

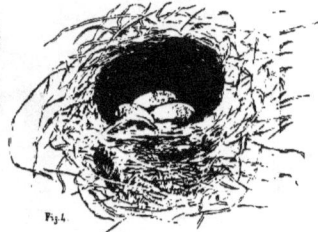

Fig. 2

Fig. 4

Nest and Eggs of
SPHENŒACUS PUNCTATUS.

Nest and Eggs of
ZOSTEROPS LATERAL
*In a branch of Leptospermum*

To accompany Paper
by
T.H. POTTS.

The Black and Pied Flycatchers breed together frequently.

Note.—Dec. 8th—Nest and eggs of *R. fuliginosa* fixed on a rock abutting on the creek in Valehead Bush, Malvern Hills; within a few feet, on the same rock, were two old nests.

## No. 47.—Platycercus Novæ Zelandiæ, Sparrm.
### Kakariki.
### Parroquet.

As far as we are aware, the breeding habits of this variety of Platycercus differ in no material point from those of *P. auriceps*. We have been told that occasionally it breeds on rocks.

Eggs, oval in shape, measure 1 inch 1½ lines in length, by 10 lines in breadth.

This species is frequently to be seen caged; in confinement it imitates the human voice, with tolerable distinctness. This bird, as well as the smaller species, is frequently shot for food.

## No. 50.—Platycercus auriceps, Kuhl.
### Kakariki.
### Parroquet.

The smaller Parroquet is a beautiful object, as with merry note it darts across the forest glade, with its bright green plumage glinting in the sunshine, giving at once a foreign impress to the scene, in the mind of the English settler.

Troglodytal in its breeding habits, it seeks some hollow tree or branch in which to rear its young; sometimes its nest is placed between the wood and the dissevered bark of a decaying tree; more frequently at the bottom of some deep hole. The eggs are white, and somewhat oval in shape.

In the gardens situated near bush, the Parroquet becomes a great purloiner of fruit. Near Arowhenua and Waimate, we have seen it rising in flocks from the oat-ricks. It is so bold as to be very easily snared with a tohe-tohe reed, noosed at the tapering point.

It commences breeding in August.

Since the great fall of snow, July, August, 1867, all bush-birds about the Malvern Hills appear to have become scarcer; for quite a year after that great storm, the silence in the bushes seemed quite remarkable, as though entirely deserted by their feathered songsters. This was notably the case in the Rockwood Bush.

## No. 51.—Nestor meridionalis, Gml.
### Kaka.
### Bush Parrot.

One of the commonest of our larger birds; yet in most of our bushes it is not nearly so numerous as it was a few years since. A troglodyte it may certainly be termed, for in the choice of a situation for its nesting-place, it seeks the shelter of a hollow tree. Sometimes the entrance-hole is a considerable distance from the nest, which is merely the decaying wood at the bottom of the hole. It lays four eggs, which, like those of most birds that breed in holes, are white; ovoiconical in form, they measure in length 1 inch 9 lines, by 1 inch 3½ lines. Sometimes, before the young are old enough to vacate the nest, it recommences laying. It is in considerable request as an article of food; they are fattest in the great Fagus forests, during the month of April. It is very easily snared, and readily tamed. The call of one bird in distress will soon cause it to be surrounded by numbers. A very common artifice is to hold a Kaka by the wing, its shrill call soon collects a crowd of its friends on the neighbouring trees, where they soon fall to the gun of the pot-hunter. During a few weeks, in July and August, 1856, Kakas were to be seen in extraordinary numbers, they were poor in condition, perhaps tamed by hunger; they appeared to have lost their wonted vivacity, and numbers were cut over with stock-whips, as they sat perched on the rail fences, about stations on the Malvern Hills. The earliest bird in the bush, its call may be heard long before daylight.

No. 58.- CHRYSOCOCCYX LUCIDUS, Gml.

Pipiwharaupa.

Bronze-winged Cuckoo, Whistler.

This beautiful little bird, in some districts, is most commonly known by the name of the Whistler. It is so called from its peculiarly clear note, which exactly resembles the sound made by a man whistling his dog.

It is remarkable for the regularity of its annual visits: in the neighbourhood of Christchurch it is almost sure to be heard about the 8th of October. We have a note of its appearing as early as the 27th of September (1855). The male bird usually selects the topmost sprays of the tallest trees for his perch, during the time it is giving utterance to its remarkable call; he seldom remains long in one spot, and indulges in a restless jerky motion of his tail. The female may be noticed very silently entering, and peering about scrubby bushes of no great height.

A parasite, like the Cuckoo of our Old Country, it saves itself all the trouble of nest-building by making use of the nests of other birds, and, of course, relieves itself of the care of providing for its young. Our experience points to the little Grey Warbler (*Gerygone assimilis*) as the most frequent victim of this "gay deceiver." The single exception we have ever observed, was the *Petroica macrocephala*, another insect-eating bird. We are not quite satisfied as to the manner in which the egg is deposited by a bird so disproportionate in size to the nest and its porch-like entrance, as that of the Gerygone. Either this Cuckoo does not destroy the eggs of the Warbler when it makes its deposit in the nest, or the bird lays to the egg of the intruder, as we have taken the nest with three of the Warbler's eggs, besides the egg of the Chrysococcyx; yet, whenever the young Cuckoo has been found in the nest, it has invariably been the sole tenant, we have not yet been able to observe in what manner the eggs, or young of the rightful owners have been extruded. In selecting the nest of so early a breeder as the Grey Warbler, as a home for its young, it secures certain advantages for the benefit of its offspring which should not be lost sight of. At the period of the Cuckoo's arrival the Warbler has most probably reared its first brood, so that even young birds have had time to gain experience in building their habitation and rearing their nestlings; then too, everywhere insect life abounds, so that a proper supply of food, sufficient for comparatively so large a bird, can be obtained by its little foster parents, with less labour and more certainty, than it could have been secured two months earlier, when several birds commence their breeding arrangements. It has been noticed, under Gerygone, why the domed nest is selected, namely, for its warmth. Then, in addition; the Golden-winged Cuckoo, be it remembered, has a most extensive range, even to the tropical islands of New Guinea, Java, and Sumatra—according to Schlegel —(see "Finsch's Notes," p. 118), and it probably chooses a pensile nest, through the same instinct, a regard to the safety of its young, which causes the *lately*-settled Zosterops to continue, for the *present*, a pensile nest-builder, though, as we have said elsewhere, we think there are indications of a change in its style of architecture. The selection, made by this Cuckoo, goes to strengthen our idea of the non-indigenous origin of our pensile nest-builders. The egg, elliptical in form, pale greenish-dun in colour, measures 9 lines in length, with a breadth of nearly 6 lines. Locality, slopes with manuka scrub, in Ohinitahi, Governor's Bay. The Whistler is a great insect eater, and appears especially fond of the well-known ladybird; we are not without certain suspicions that it devours or destroys the eggs of other birds.

### No. 60.—Coturnix Novæ Zelandiæ, Quoy.
#### Koreka.
#### Quail.

This excellent game bird is almost extinct, but a few years since it existed in the utmost abundance; bush fires, extending often for many miles, must have been the active agent in destroying a bird possessing such limited powers of flight, as our handsome little Quail.

A very slight nest, composed of a few bents of grass twisted into a depression of the ground, was all the artificial shelter this bird relied on, for the purpose of incubation. The eggs were very numerous; we have been told that as many as ten or twelve have been found in a nest, oval in shape, colour buffy-white suffused with rich brown splashes, with a remarkably glossy varnish; length 1 inch 3 lines, by 11 lines in diameter. We have not heard its call-note, or seen a bevy of Quail, for years. The sheltered valleys round Lake Coleridge, and about the head-waters of the Rakaia, were the last places in which it lingered, to our knowledge. They bred more than once in the season, as we have a note of abundance of young Quail so late as the 9th and 10th of April (this was in 1857). We have seen it escape the talons of the Quail-hawk, by dropping perpendicularly, just when about to be struck, when all hope of escape from its relentless pursuer was quite abandoned. The flight of the Quail is low, and it used to be said that it would not rise after being flushed the third time: numbers were killed by sheep and cattle-dogs in the early days, when it abounded. In style of flight, our bird must resemble the Quail of Taberah and Kibroth-hattaavah, that fed the Children of Israel, in the wilderness:—"And there went forth a wind from the Lord and brought quails from the sea, and let them fall by the camp, as it were a day's journey on this side, and as it were a day's journey on the other side, round about the camp, and as it were *two cubits high* upon the face of the earth." Our bird is not migratory as we believe. The young, with the exception, perhaps, of that of *Apteryx Owenii*, undergoes less change in plumage than that of any other bird; the young, when it assumes its feathers, exactly resembles the adult female, with the white streak along the shaft of the feathers, which adds so much to its beauty.

### No. 61.—Apteryx Australis.
#### Kiwi.

We have not enjoyed an opportunity of acquiring, from personal observation, any knowledge of the breeding-habits of the curious family of Apterygidæ: a description of the eggs of the different species may be thought not out of place in the present paper. We believe this species is peculiar to the Middle Island.

An egg received at the Canterbury Museum from Okarito, or its neighbourhood, is believed to be an undoubted specimen of this species,—it arrived, in a fresh state, in November. It was white, much blunted at each end, and presenting a very smooth surface: this enormous egg gives the following measurements: through the axis 5 inches 1 line, with a breadth of 3 inches 4 lines.

Rev. J. G. Wood, in his "Nat. Hist. Birds," writes of the eggs laid by the Kiwi at the Zoological Gardens, London: "These eggs are indeed wonderful, for the bird weighs just a little more than four pounds, and each egg weighs between fourteen and fifteen ounces; its length being $4\frac{3}{4}$ inches, and its width rather more than 2 inches."

The Canterbury Museum also contains some fine specimens of the bird, obtained from Westland.

## No. 62.—APTERYX OWENII.
### Kiwi.

The smallest and most common of the whole family. Specimens of Owen's Apteryx are not very rare in collections, but the celebrity which attaches to this wingless genus is rapidly drawing down destruction upon it. No mercy is shown to it, and there is no exaggeration in stating that a regular trade is carried on in specimens of these birds, and the equally unfortunate Kakapo (*Strigops habroptilus*). Could not our paternal Government interfere in behalf of these interesting aborigines, for we believe there are those who would shoot the Cherubim for specimens, without the slightest remorse. This species is peculiar to the Middle Island.

An egg in the Canterbury Museum, from the West Coast, measures 4 inches 6 lines in length, with a breadth of 2 inches 7 lines (other specimens we have measured are of rather larger dimensions); colour, white, with a very smooth surface, blunt at each end.

The young of the Kiwi, without exhibiting any sign of an immature state of plumage, is disclosed, as it were, from the shell, arrayed in the hair-like integument of an adult bird. In this species the mottled-grey feathers of old and young appear to be of same shade of colour.

## No. 63.—APTERYX MANTELLII.
### Kiwi.

This is usually known as the Kiwi of the North Island, and it is believed that it has become comparatively rare during the last few years.

An egg, in our own collection, from Whangaroa, measures not less than 5 inches 4 lines in length, with a breadth of 3 inches 3 lines. This specimen is white, of smooth surface, rather more pointed at one end than is usually the case with the eggs of this family.

## No. 65.—CHARADRIUS BICINCTUS.
### Banded Dotterel.

The family of the Charadriæ have always been distinguished for their wariness, and the artful devices employed to allure strangers from their nest. Heliodorus gives such a singular reason for its shyness, that we cannot resist quoting it: "The bird Charadrius cures those who are afflicted with the jaundice. If it perceives, at a distance, any one coming towards it, who labours under this distemper, it immediately runs away, and shuts its eyes; not out of an envious refusal of its assistance, as some suppose, but because it knows by instinct, that, on the view of the afflicted person, the disorder will pass from him to itself, and therefore it is solicitous to avoid encountering his eyes." Our banded Dotterel is worthy of belonging to the family of the Charadriæ, for it is one of the most restless and wariest of birds, during the breeding season. On the approach of an intruder, it flies round and round, uttering its note of warning, then alighting on some rising ground, it steadily keeps watch. During the time it remains on the look out, it indulges in a peculiar habit of jerking its head backwards and forwards, uttering its monotonous twit, twit, at intervals.

It commences breeding early in the spring; its simple nesting-place may be found on " the plains," or in river beds. It lays three oval-shaped eggs,

greenish-brown, much sprinkled with dark-brown markings; they measure
1 inch 4 lines in length, with a breadth of 1 inch.

The young are exceedingly active, the little brown puffs of down may be
observed running with great swiftness on being alarmed. In the autumn the
Dotterel assembles in flocks of considerable numbers.

NOTES.—August 2, 1856, saw a nest and two eggs. Rakaia river.
September 1, 1856, saw nest and three eggs, Rakaia river.
October 14, 1857, young birds quite strong.

### No. A. 65.—CHARADRIUS OBSCURUS, Gml.
### Tituriwhatu-pukunui.
### Red-breasted Plover.

In Dr. Finsch's list, in Vol. i., "Transactions New Zealand Institute,"
this Plover is named *Hæmatopus obscurus.*

This handsome bird is to be met with on hill and plain, yet nowhere in
very considerable numbers. In the breeding season we have noticed it at such
a considerable altitude as the summit of Dog Range, in the Ashburton district.
The nest is difficult to find, it is so slight an affair that it easily escapes observation, merely a few stems of grass twisted into a slight hollow in the ground,
so loosely put together that it is not easy to pick it up and yet preserve its
form. The eggs, three in number, just fill the nest; they are of a delicate
soft-brown, suffused with dark-brown, almost black, marks, somewhat oval in
shape, in length 1 inch 9 lines, with a breadth of 1 inch 3 lines. The young
run with speed almost as soon as hatched, and conceal themselves with much
skill. Young birds have not the rufous tinge on the breast and upper part of
the abdomen. We have observed eggs and young in the months of October
and November.

NOTE.—Oct. 22, 1867—Nest with three eggs;—saw young Plovers.

The warning-note of this bird sounds like click, click, slowly repeated.

An excellent figure of it, rather warmly coloured, is to be found in Ross's
"Voyage of the Erebus and Terror," Vol. i., plate 9, Birds.

### No. B. 65.—ANARHYNCHUS FRONTALIS, Quoy. and Gaim.
### Scissor-bill,
### Crook-billed Plover,

Appears in Dr. Finsch's list, "Transactions New Zealand Institute," Vol. i., as
*Hæmatopus frontalis.*

The Crook-billed Plover, at the breeding season, is less wary than any
of its congeners, and its nesting-place would be discovered with very little
difficulty, were it not for the wonderful instinct it exhibits in selecting the ground
for depositing its eggs. They are simply laid, without any preparation, amongst
the pebbles of some river-bed usually, and never far from water, and so well
does their grey tint harmonize with the general colour of the shingle around
them, that their detection would be almost hopeless if this bird was less
confident.

Its oval-shaped eggs are three in number, grey stone-colour, with the
whole surface minutely dotted over with black specks; they measure 1 inch
4½ lines in length, with a width of 1 inch ¼ line. On approaching the eggs or
young, the old bird trots slowly away, assuming a broader and somewhat flatter
appearance, by slightly extending the wings, making at the same time a low
purring sound.

Breeding season extends from September to December.

The young birds are covered with grey down, and appear to have legs
long, out of all proportion to the size of the body; at this early stage, the
peculiar deflection of the bill, although slight, is perceptible; it is always

turned to the right, or off side. Birds of the year, we believe, do not assume the frontlet which distinguishes the old birds, and which is broadest in the male. No satisfactory reason has been given for the peculiar form of the bill of this bird, which exceeds in length that of *C. bicinctus.*

NOTES.—Sept. 14, 1856—Saw three eggs on a patch of small shingle.
Oct. 28, 1857—Young birds on the Rakaia river-bed.
Oct. 30, 1867—Two eggs on the bare shingle, Rakaia river.
Nov. 2, 1867—Three eggs chipped, on shingle, Ashburton river.

### No. 71.—HÆMATOPUS LONGIROSTRIS, Vieil.
### Torea.
### Oyster-catcher, Red-bill.

The Oyster-catcher is one of the wariest and most restless of our birds, ever ready with its clamorous alarm-note, to wake up each echo, and disturb every bird within the sound of its shrill cry; but in the breeding-season it exhibits an intensity of slyness, that is almost supernatural. Usually it breeds in our river-beds, on the sandy spits, without other shelter than what may be afforded by some drift flax, grass, or stick, near which it makes, or discovers, a slight depression, in which to deposit its eggs, which are somewhat oval in shape, 2 inches 3 lines in length, with a diameter of 1 inch 7½ lines; pale or yellowish-brown: these are not to be distinguished from those of the European bird, much covered with irregular marks and spots of rich brown. Usually three eggs are laid, but we have found it incubating a single egg: the young are grey, with a dark longitudinal stripe on each side, above the wing. They are very active, and are early led by the old birds to the margin of the water-holes or pools. On being alarmed, the old bird sidles off the nest quietly, takes advantage of any broken ground that apparently conceals its movements from observation, and makes a long detour; a close scrutiny will very frequently enable the observer to detect the head of the bird carefully peering out behind some vantage-ground, watching all his proceedings.

A very common frequenter of the coast, as its familiar name imports; in the winter time it assembles in large flocks on the mud flats disclosed by the ebbing tide; though a shore-bird, it is found breeding in solitary couples, often far inland, certainly sixty or seventy miles from the sea, for instance, up the Wilberforce river, nearly as far back as the neighbourhood of Browning's Pass. A pair will boldly attack the Harrier, male and female striking at the Hawk in turn, and driving it to a safe distance from their young. Hæmatopus, that is, literally, the blood-red foot, one of the birds mentioned by Pliny, appears to be universally met with.

### No. 75.—BOTAURUS POICILOPTERUS, Wagl.
### Matukuhurepo.
### Bittern.

Not so frequently met with as before such an extensive breadth of swamp-land had been drained and cultivated. It was once very common about Christchurch, "the City of the Plains," it still haunts the banks of the Avon, and breeds in the neighbouring swamps. The breeding season of the Bittern must extend over a considerable period, as we have found the eggs, quite fresh, in the middle of January (15th). A nest near Clearwater, or Lake Tripp, in the Ashburton country, was built of raupo, (*Typha angustifolia*), surrounded by water about ankle deep; the top of the nest was very flat, and stood about six inches above the surface of the water. (See Plate 4, Fig. 7.) We have not seen more than four eggs to a nest, they are oval in form, varying slightly in colour, from buffy-brown to pale olive-green. Through the axis they measure 2 inches 1½ lines, with a diameter of 1 inch 6 lines.

### No. 78.—HIMANTOPUS NOVÆ ZELANDIÆ, Gould.
#### Poaka.
#### Pied Stilt.

Usually commences breeding in October, according to our experience. Unlike the black variety, we have always found this bird prefers swampy ground, such as fringes the shores of shallow lagoons, as a nesting-place, etc. Here it may be observed busily wading in the shallow water. In its habits of nidification, it is as inartificial as its congener. Eggs, four in number, yellowish-brown in colour, about the same size as those of the Black Stilt, are profusely marked with very dark brown; we have thought the eggs rather warmer in colour than those of the black variety. The monotonous call of pink, pink, has, in some places, fixed on it the trivial name of Pink.

### No. B. 78.—HIMANTOPUS MELAS, Homb. et Jacq., "Ann. des Sci. Nat., 1841."
#### Black Stilt.

Breeds early in the season, seeking the sandy river-beds for that purpose. The labour of nidification is very trifling, sometimes a nest of grass, etc., is roughly constructed, now and then this apology for a nest may be discovered on a log of drift-wood; much more frequently, however, a slight depression in the sandy spit, answers all the requirements of this Stilt, as a nesting-place; it is never very far from water.

It lays from three to four eggs, rather oval in shape, yellowish-brown in colour, very much spotted and blotched over with very dark-brown, approaching to black, measuring 1 inch 10 lines in length, by 1 inch 3 lines in breadth. The young can run almost as soon as they are hatched; when disturbed, they conceal themselves behind stones, or some other shelter, in the most artful manner; they are covered with dark-brown down, bills and legs are then very dark, almost black. The parent birds exhibit the utmost assiduity in attempting to lead intruders from their eggs or young, and their numerous cunning devices are carried on with surprising cleverness and perseverance.

We have been told that there is not a Black Stilt, that the Black Stilt, so called, is, in reality, but the pied species in an immature state of plumage. To this we cannot for a moment subscribe, we have never once found the two species breeding together or using the same, or even similar situations, for their nesting-place.

Neither Mr. Buller nor Dr. Finsch, we remark, admit this bird in their lists, but, with the utmost deference to those authorities, we cannot consent to give up such an old acquaintance as the Black Stilt. Our opinion on this subject, is shared in by many others, living "up country," who have had good opportunities, for several years, of observing the breeding habits, and the young birds of both species of Stilts.

NOTES.—Sept. 13—Nest with three eggs, on a spit on Rakaia river-bed.
Dec. 14—Nest with two eggs, on a drift-log in the Rakaia river-bed.

### No. 87.—OCYDROMUS AUSTRALIS, Sparrm.
#### Weka.
#### Wood-hen.

This bird is so mischievous to the fruit garden and poultry-yard of the up-country settler, that unrelenting war is usually waged against it. Small fruit, low-growing apples, eggs, and young chickens, form some of the items of its favourite plunder; nor is its thieving propensities confined to articles of food, spoons, pipes, pannikins, and a long list of miscellaneous articles, we have known this *curious collector* to carry off. On one occasion, in Alford Forest, a watch was stolen, and accidentally recovered, a few days after, at some little distance from the hut ; for the Weka, unlike the Jackdaw at home, does not appear to care for a secret hiding-place in which to conceal its pilferings.

We have seen it kill a well-grown Spanish chicken, six weeks old, with one blow of its powerful bill. Some time since, a Weka appeared in our garden, much to our gratification, for, in the neighbourhood, the bird was of very rare occurrence; all went well till the first brood of choice Dorking chickens was discovered, and then, well, the Weka had to die.

At night, and before rain, the loud screaming of this bird is most frequently heard. The nest is found in a variety of situations, such as in a tuft of Celmisia, grass-tussock, or sometimes in a thicket of young plants, on the outskirts of the bush ; we have observed it under the shelter of a rock, without any attempt at concealment, which the tussocks growing close by would have afforded. Grass is usually the staple material of its home, which is large, and basin-shaped within. The eggs, from five to seven in number, are white, with reddish marks generally distributed over the surface ; but in many specimens the colouring is most abundant at the larger end. The young, covered with very dark down, may be observed, like chickens, following the old bird, who collects them around with the call of toom, toom, repeated quickly, and much lower in tone than the booming note to which the Weka sometimes gives utterance, and which is probably the call of the male. As the young grow up, the dark-brown of its early days gives place to a more mottled plumage when about one-third grown ; although the legs become lighter in colour, the beak still retains its dark appearance. There is much difference in the size of Wekas, some of the hill-birds are very large, and we expect that before very long they will be classed as a separate species.

A very light-coloured specimen was observed near Mount Hutt, last year. Numbers of these birds are killed for their oil, which is much esteemed by bush-men for a variety of purposes ; properly dressed, they are excellent as an article of food, due care being taken as to where they are obtained, as they are very foul feeders.

No. 91.—PORPHYRIO MELANOTUS, Temm.<br>Pukeko.<br>Swamp-hen.

This beautiful rail delights in swamps, where its nest is also to be found, built of grass ; the top is sometimes more than a foot above the ground, and not unfrequently it may be observed surrounded by water. The number of eggs to a nest varies considerably, as we have found from two to seven, five may be considered the usual complement, in shape ovoiconical, greyish-brown, with dots and blotches of reddish-brown, measuring through the axis 2 inches 2 lines, with a diameter of 1 inch 6 lines. These dimensions appear very small for so large a bird, more especially when compared with those of the egg of *Apteryx Mantelli*. The young run about as soon as they are hatched, and on being disturbed conceal themselves with great art. They are thickly clothed with black velvety down, interspersed with fine hair-like points of silver-grey ; legs dullish-red, beak has a yellowish ivory look, which contrasts pleasingly with the rest of the body. The Pukeko is esteemed excellent eating.

No. 92.—CASARCA VARIEGATA, Gml.<br>Putangitangi.<br>Paradise Duck.

This well-known bird often chooses the shelter of a huge tussock, beneath which to make its nest ; sometimes a hole in a rock is chosen in preference.

Nest of
CERYGONE ASSIMILIS.
Front and Porch ornamented with
burrs of Acæna. (side view.)

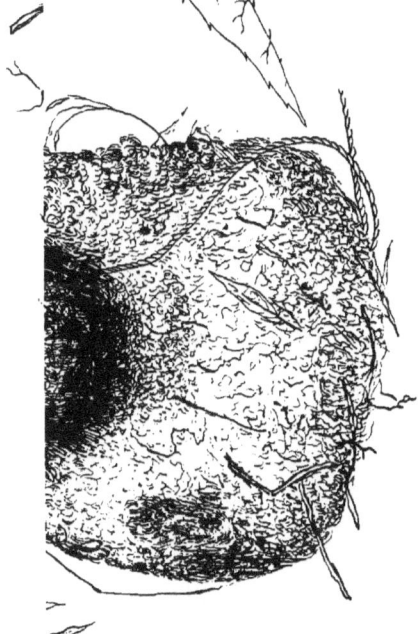

Nest of
CERYGONE ASSIMILIS.
From the fork of a willow tree.

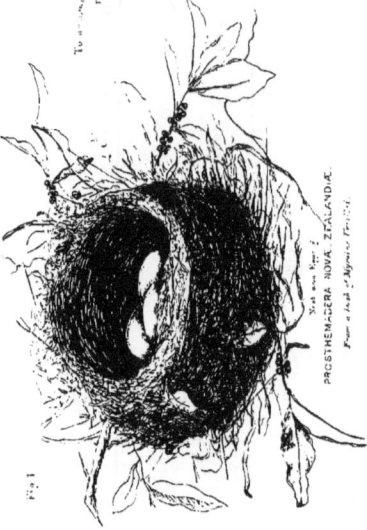

Fig. 1

To accompany Papers by W. W. Smith

PROSTHEMADERA NOVÆ ZEALANDIÆ.

From a Sketch by Keulemans Frohawk.

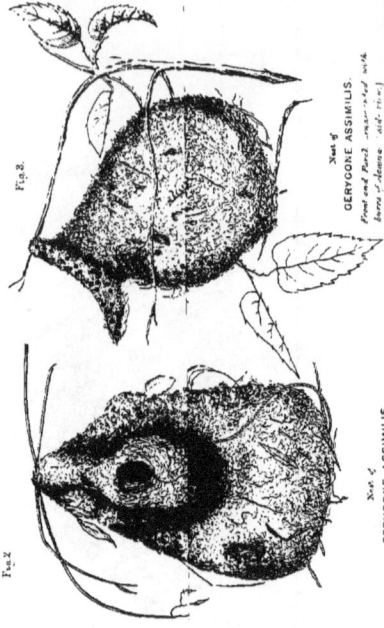

Fig. 2

Nest of
GERYGONE ASSIMILIS.

Fig. 3

Nest of
GERYGONE ASSIMILIS.

Front and Back; suspended with berry of Solanum aviculare.

We know a large rock, on the bank of the Rakaia, where a pair of these birds breed every year.

The nest is warmly lined with down ; nine eggs are sometimes found in a nest, but not often have we noticed so large a number ; they are large, creamy-white, ovoiconical, vary somewhat in size, even in the same nest ; length 2 inches 9 lines, with a diameter of 1 inch 10 lines. The Paradise Duck leads its brood to water very soon after hatching. The parent birds may be noticed surrounded by their tiny young ones, spending nearly the whole day upon the water, even when the usually smooth surface of the lake has been lashed into foam-crested waves by a furious Nor'-wester. They enter the lake after the sun is well up, and remain till late in the afternoon ; this is daily repeated, the young birds gradually venturing farther from the old ones, and may be observed darting about with the greatest activity.

This bird employs the wiliest stratagems to lead the wayfarer from its nest or young.

NOTES.—October 24, 1855, noticed nest with eight eggs, Malvern Hills.

November 1, 1867, saw nest with five eggs, another with six eggs, on the Potts river.

December 2, 1867, Duck sitting on five eggs, Rangitata river.

The congress of the sexes takes place in water, after the manner of the common Goose. When young, the flesh of the Paradise Duck is very good eating, but in old birds there is a degree of toughness, that only the sharpest appetite can overcome.

The young are easily tamed, and feed amicably with other poultry ; but unless confined when spring sets in, they are almost certain to ramble away and be lost.

## No. 93.—ANAS SUPERCILIOSA, Gml.
### Parera.
### Grey Duck.

One of the commonest game-birds left to us by the eager sportsman. We have found the nest of the Grey Duck in so many situations, differing so entirely in character, that it would be difficult to pronounce any one position as the favourite site for its breeding-place. Sometimes close by the edge of a bush creek, amongst damp shady ferns ; out on the plain, sheltered by a tussock, quite away from water ; often on a hill side. Yet, whether on the level plain or in a swamp, its cup-shaped nest is most profusely lined with down, and diffuses a strong musky odour. The eggs, usually nine in number, are creamy-white, occasionally varying a little in size and shape, some are ovoiconical, others broadly oval ; 2 inches 6 lines through the axis, with a diameter of 1 inch 8½ lines, is the measurement of a large specimen ; whilst we possess specimens that measure but 2 inches 3 lines in length, with a diameter of 1 inch 7 lines. On referring to many notes on the numbers of eggs laid by the Grey Duck, an entry appears of ten eggs in a nest, found December 10th, near a lagoon by the Rangitata, the largest number of which we have a memorandum. We have seen the young quite tame, and associating with the common domestic Duck.

## No. 94.—ANAS CHLOROTIS, Gray.
### Puteke.
### Teal.

Very much scarcer of late years than we can remember it. A few years ago we used to hear tales of almost incredible bags of this excellent game-bird ; a few more years of inconsiderate slaughter, will make the Teal a rarity.

E

The nest is made of grass, thickly lined with down, sometimes close to the edge of a swampy creek, or beneath the sheltering leaves of a large "Maori-head" (*Carex virgata*).

The eggs are large for the size of the bird, cream colour, not unlike those of the Mountain Duck, in tint, but perhaps slightly darker; length 2 inches 5 lines, diameter 1 inch 10 lines, We have not found more than eight eggs to a nest. On a pond at Rockwood, in the Malvern Hills, three Teal fraternised with some tame Paradise Ducks, and came regularly, to be fed, every day, with pieces of bread.

### No. 96.—FULIGULA NOVÆ ZELANDIÆ, Gml.
#### Papango.
#### Black Widgeon.

In the hill-country, a few years since, this was sufficiently common; a small tarn, near Lake Coleridge, yet retains the name of Widgeon Lake, from the numbers which formerly frequented it. Near one small pool in the Ashburton country, where it bred in considerable numbers, neither birds nor nests are now to be met with.

A gregarious bird, it delights to assemble in large flocks, and may be seen on some of the more secluded lakes, swimming about, and disporting with numbers of other water-fowl, very frequently diving. Sometimes it breeds in the shelter of a huge "Maori-head." We have found it well concealed by a large snow-grass tussock, within a few feet of water, where there was a rent or crack in the ground. Nest of grass, thickly lined with down, contained five eggs of a deep cream-colour, ovoiconical in form, measuring 2 inches 8 lines, with a diameter of 1 inch nearly 9 lines.

### No. 98.—HYMENOLAIMUS MALACORHYNCHUS, Gml.
#### Whio.
#### Blue Duck, Mountain Duck.

The only way of seeing this singular bird to advantage, is by paying a visit to the mountainous districts. On a mountain torrent, where the foaming water dashes from rock to rock in countless eddies, the Mountain Duck lives at ease, making its way up or down stream. Sometimes it may be observed basking in the sunshine, near a shallow pool of the rapid streamlet. Sometimes it is a burrower, and its nest may be found in a hole in a bank; we have found it concealed from view by overhanging sprays of those various alpine Veronicas, which sometimes make the mountain creeks in the back-country perfect gems of beauty. The nest, like that of other ducks, thickly lined with down, we have found to contain five eggs, of a deep-cream colour, elliptical in form, measuring 2 inches 8½ lines in length, with a diameter of 1 inch 9 lines.

One of our early breeders; we have known the young brood to be swimming about by the end of September. We have seen nests of eggs in October and November. A much frequented breeding-place is above the gorge of the Potts river,—a tributary of the Rangitata.

### No. 99.—PODICEPS RUFIPECTUS, Gray.
#### Totokipio.
#### Dab-chick, Little Grebe.

This bird is far from uncommon, and is to be met with on lakes, lagoons, and deep creeks that run still and swift, unlike the noisy torrents in which the Mountain Duck delights.

The nest is rather a large and somewhat clumsy structure, formed of the roots and leaves of various aquatic plants. We have found it built against the stem of the *Carex virgata*, beneath the drooping leaves of which it was perfectly concealed from casual observation. Situated just within the swampy side of a small lake, it was raised a few inches, only, above the water-level. We have invariably found two eggs to a nest; they are greenish-white, frequently with wart-like protuberances, and more or less weed-stained.

Eggs from the same nest occasionally differ a little in size, as may be observed from the following measurements: length 1 inch 9 lines, by a diameter of 1 inch; whilst another egg, from the same nest, measured in length 1 inch 6½ lines, with the same diameter as in the longer specimen. The lobed foot of the Grebe is a remarkable peculiarity, assisting it to swim and dive with great rapidity; in its habits it appears much more restless and fidgetty than the large Crested Grebe. The young is greyish-brown on the back, warm-rufous on the neck and breast, lighter on the abdomen; the head is beautifully mottled with black, and rich reddish-brown alternately. When alarmed on the water the parent birds have a knack of tucking the young under the wing, so that its head is alone visible; they dive and swim, thus encumbered, with the greatest ease.

### No. 100.—Podiceps Hectori, Buller.
#### Crested Grebe.

One of the most ornamental of the water-fowl, that add so much to the interest of the lake scenery of our Southern Alps. In April, 1856, we first made its acquaintance, on a small lake, now called Lake Selfe. It appears to move about in pairs, as a single couple is usually found (or rather was to be found) associating with nearly every group of Ducks that dotted the little secluded bays of the lakes.

The Grebe swims low in the water, with a certain air of demure gravity, which affords a marked contrast to the rapid movements of most of the other natatorial birds, with which it so frequently associates.

We have found the nest in November and December. The structure is large, and very solidly built of pieces of decayed *Carex virgata*, raised about a foot above the level of the water: its sloping sides give a ready means of reaching the basin-like depression on the top, in which the eggs are deposited. (See Plate 4, Fig. 5.) In several instances we have observed that the nest had been constructed on the top of an old stump of *Carex virgata*, situated in a shallow part of a lake, distant from twenty to one hundred yards from the shore. Last Spring, in the little boat-harbour on Lake Coleridge, belonging to Mr. Oakden, there happened to have been thrown a quantity of cut flax, which the bight of a chain prevented from drifting out to the lake; a pair of Crested Grebes built on this floating mass, and succeeded in rearing their young; it should be added, this harbour is not much used, and the proprietor is a careful protector of our native birds. We have known three instances, in which the nests have been submerged by the rising waters of the lake, an indication that such mishaps must frequently occur, which may perhaps in some measure account for the comparative rareness of this Grebe.

The eggs, three in number, are at first greenish-white, but very quickly become stained over, entirely, with yellowish-brown, from the water and weedy slime adhering to this bird's flat lobed foot. Eggs elliptical in form, measuring through the axis 2 inches 4 lines, with a breadth of 1 inch and nearly 7 lines. We believe that an interval of two, perhaps three, days occurs between the laying of each egg. The young bird is pale-brown with dark brown marks. During incubation the Grebe maintains an upright posture, with its long graceful neck held erect, so motionless its attitude, that at a distance it rather resembles a stick than anything endued with life. Watchful and shy, it noiselessly dives, immediately it discovers itself observed. The power of diving, and the faculty of remaining under water for a considerable time, is too remarkable a characteristic of this bird to be passed over without notice.

### No. 104.—Spheniscus minor, Forst.
### Korora.
### Small Penguin.

One of our commonest sea-fowl; and certainly a frequent burrower in its mode of nidification.

We have found the Penguin breeding every year, in the inner chamber of a deep cave, perfectly dark ; a hollow, scraped out of the sandy bottom of the cavern, half filled with fish bones, formed the nesting-place, in which the eggs were deposited. We have always found two eggs, as the complement of the Little Penguin.

The eggs are white and very smooth, but soon become stained ; they are broadly oval, and measure, through the axis, 2 inches 2 lines, with a diameter of 1 inch 9 lines. The old birds defend their nests with great spirit, using beak and claws most vigorously, making at the same time a noise not unlike the mewing of a cat.

We have observed these birds breeding during the months of November, December, and January. They breed in great numbers amongst crevices of rocks ; in sand-banks, their tunnels are bored with great neatness, with a run frequently extending a considerable distance. The entrance generally exhibits a perfectly round hole, about three or four inches in diameter, and from whence is diffused a most powerful odour. The young, we have found in the nest when nearly full-grown ; their slatey-bluish plumage is brighter than that of the parent birds. We have an egg, very much encrusted, showing a departure from their usual appearance, which is usually as white and smooth as that of the domestic fowl. In retreating to the sea, its action is peculiar, walking it can scarce be called, it appears to throw the whole body forward, and shuffles along with an undulating motion, which gives the Penguin more the appearance of a large grey rat than that of a bird.

### No. 126.—Larus dominicanus, Licht.
### Kororo.
### Grey Gull, Black-backed Gull, Large Gull.

Our larger Gull breeds on the sea-shore, upon the sandy spits in the river-beds. The rough-looking nest is large, usually made of grass, sometimes of small tussocks pulled up by the roots. We have noticed these birds visiting the breeding-ground early in August, but have not seen the eggs till some weeks later, apparently these must have been visits of inspection, when they busied themselves about the nests in rather a clamorous manner for several days in succession. The eggs, two or three to a nest, are ovoiconical, measuring 2 inches 10 lines in length, by 1 inch 10 lines in width. The colour varies from shades of light-grey to brown, covered, more or less, with grey and brown marks and blotches. The young are well covered with grey down at first, they assume, gradually, a mottled-brown plumage, the bill still dark. presenting a marked contrast to adult birds. The parent birds defend their nest with great spirit, a pair will drive away, and give chase to, a Harrier. Their olfactory organ must be most acute, as they find out the carcase of a dead sheep or bullock with great readiness.

We have often been amused by watching their grotesque action in following a retreating wave, where the sea has rolled in heavily on the inclined sandy beach. A number of these Gulls wait till the wave has just expended its force and follow the retreating waters rapidly, by a series of hopping jumps, feeding the while, and sometimes only just avoiding the next incoming wave, by taking wing for a few yards with apparent reluctance.

### No. 127.—LARUS SCOPULINUS, Forst.
### Tara-punga.
### Little Gull.

During the breeding season, our very pretty Little Gull frequents the river-beds, and shores of lakes, in very large numbers. It deposits its eggs with scarcely any of the preparation that distinguishes the larger species of Gull. The eggs are usually found on the bare ground ; at most a few bents of grass, amongst the stones, sufficing for a nest. The eggs, often broadly oval, sometimes ovoiconical, are of different shades of greyish-brown, plentifully besprinkled with darker marks and blotches of grey and brown. Length, 2 inches 1 line, by a diameter of 1 inch 6 lines.

### No. 129.—STERNA CASPIA, Pall.
### Fish-hawk.

This fine Tern is content with merely a hollow scraped in the sand, just large enough to contain the eggs ; the breeding season extending from November to January ; our earliest note of having seen the egg, is dated November 14th.

The eggs, usually two or three in number, ovoiconical in form, measure 2 inches 7 lines in length, with a diameter of 1 inch 9 lines ; we have a specimen from Lake Ellesmere, much smaller than is shown by this measurement ; the eggs are of varying shades of pale greyish-brown, richly spotted with dark-grey and brown, distributed all over the surface ; in some specimens these markings are most numerous at the larger end. When these birds are disturbed at breeding-time, they ascend to a great height, and hover around the intruder, uttering loud screams. We have found the young as large as the adult *Larus scopulinus*, before they were able to fly. Have found this bird incubating a single egg.

### No. 130.—STERNA LONGIPENNIS, Nordm.
### Whale-bird.

The black-billed, swallow-tailed Whale-bird seems constantly to frequent our coasts and harbours, the liveliness of its movements on the wing, especially the rapidity with which it drops from a great height to secure its finny prey, frequently renders it an object of remark to the dwellers on the sea-shore, it deposits its egg on the bare rock, without the slightest protection, at a distance varying from about five to six feet and upwards from the level of high tide ; the egg must often lie within reach of heavy showers of spray. Ovoiconical in form, generally, but sometimes rather oval, the egg measures 1 inch 10 lines in length, with a diameter of 1 inch 4 lines. Colour varies from shades of pale-grey, sea-green, stone-colour, or light-brown, lightly freckled with brown, or profusely blotched with slatey-grey, and chesnut-brown, to almost black. The young covered with mottled-grey down, varying in shade to almost brown, are quite helpless for two or three weeks after hatching, and appear quite unable to attempt securing safety by swimming, like young Gulls, when alarmed ; they retain the grey feathers on the head even when well-grown. Great quantities of small fish may usually be noticed surrounding the young birds. We believe this bird lays but one egg, but are aware that others entertain a different opinion. On a rocky point, in Port Cooper, which is washed with abundant showers of spray under a strong N.E. breeze, we observed about 200 birds breeding; except in three cases only, the eggs were solitary.

NOTE.—Dec. 14—Found two eggs lying together, differing in size and colour so much, that there is not much doubt they were the produce of different birds.

## No. 131.—STERNA ANTARCTICA, Forst.
### Common Tern.

In this paper on our Birds, the nomenclature followed is that which is given in Dr. Otto Finsch's Notes, "Trans. New Zealand Institute," Vol. i., pp. 122-5, but in the case of this bird we prefer adhering to the name assigned to it by Forster. In a note in the volume referred to, page 121, *S. antarctica*, Forst., is asserted to be the same species as *S. minuta*, Linn. Mr. Buller, in his "Notes on Herr Finsch's Review," tacitly admits this by his silence ; we think this must be an error. Yarrell, in his "History of British Birds," Vol. iii., p. 525, writes of *S. minuta*, "their eggs are of a stone-colour, spotted and speckled with ash-grey and dark chesnut-brown, the length 1 inch 4 lines, by 11 lines in breadth." This measurement is exactly the size of the eggs of the next species, which we have numbered A. 131, whereas the eggs of the yellow-billed *S. antarctica* measure in length 1 inch 6 lines, by a breadth of 1 inch 1½ lines, and present a very striking contrast in colour ; they differ also in shape. On referring to our collections of British and New Zealand eggs, and comparing the eggs of these species of Terns, any hesitation we may have entertained about the correctness of adhering to Forster, instead of the more modern authorities, is removed. The Common Tern, very often termed the Whale-bird, seems even more gregarious than its congener *S. longipennis*, that is, taking into consideration its habits throughout the year. It may be observed hovering over the newly-ploughed fields in great numbers, in search of larvæ of various insects ; the small lizard seems a favourite morsel, and may frequently be noticed dangling from the beak of this Tern.

We have remarked, several times, a curious habit of this bird, which presents a singular appearance to the gaze of the traveller : a large flock will rest motionless on the ground, with their delicate bluish-grey wings extended vertically, and will maintain this singular posture for some time. It deposits its eggs, two in number, on the bare ground, without any attempt at nidification ; ovoiconical in form, they measure in length 1 inch 6 lines, with a breadth of 1 inch 1½ lines. In colour very considerable variety is exhibited, dull grey, greenish-white, pale-green, pale-brown, with small brown markings distributed over the surface generally. This Tern is remarkably clamorous at the breeding season ; and should a traveller approach their ground, the intruder is instantly assailed by them with swift dartings, accompanied by noisy, harsh, grating screams. The young birds remain about the breeding-ground for some weeks, till they can fly well.

## No. A. 131.—STERNA——(?) New Sp.

A very small Tern visits the Rakaia river-bed during the breeding season, not far below the gorge of that great river. There does not appear to exist any description of it, either in Mr. Buller's Essay, Dr. Finsch's Notes, or in Mr. Buller's Paper of August 25, 1868. It lays its eggs, two in number, on the bare ground, they are stone-colour, blotched over with grey markings, and measure through the axis 1 inch 4 lines, with a diameter of 11 lines. It is not at all a common bird in that locality, and was not observed there last year. In the Museum at Christchurch, are two specimens of a small Tern, obtained in the province ; in all probability, the eggs noticed above, belong to individuals of this species. They are labelled *Sternula nereis*, and measure, total length 10 inches 7½ lines, length of wing 9 inches 1½ lines, bill from gape 1 inch 9 lines, tarsus 7½ lines; colour, above, silver or French-grey, forehead white, back of head and nape of neck black, black streak round the eye, bill and feet yellow: the eggs above noticed were discovered in November.

Since the above was written, through the zeal of a friend residing near the Rakaia, we have received two eggs of this interesting bird; they were found in November, hard set. On comparing them with the egg of *S. minuta*, of Europe, in our own collection, we find them of rather a broader oval in shape, of the same length, with a breadth exceeding that of the European species by 1½ lines; but so close is the general resemblance between them, that they might be easily supposed the produce of individuals of the same species. The egg of the Lesser Tern, *S. minuta*, is less eccentric in its colour and marks than those of many other species of the genus.

### No. 139.—GRACULUS BREVIROSTRIS, Gould.
### Black River-shag.

Breeds on the shores of the lakes in the interior, where these birds congregate in considerable numbers, probably depending on the fresh water Unio, for some portion of their food supply. Like the Rook, and the Heron, of Europe, it builds in company, within the space of a few square yards many nests may be counted; the favourite breeding-place appears to be scrub, on some of the steep and lofty banks of the lake shore. The nest is large, chiefly constructed with sticks procured from the dead scrub, amongst which may be found the dead flower-stalks of *Aciphylla squarrosa*, grass forming the lining material. The eggs, four in number, are greenish-white, with the chalky encrustations characteristic of the Pelicanidæ, elliptically shaped, they vary considerably in size, especially in the measurement through the axis, as from 2 inches 6 lines, to 2 inches 2½ lines, with a breadth of 1 inch 6 lines. When freshly procured from the nest they give out that peculiar odour which distinguishes those of the Procellaridæ, in common with the eggs of the Pelicanidæ, truly "a most ancient and fish-like smell."

### No. 142.—DYSPORUS SERRATOR, Banks.
### Gannet.

An egg of this bird, in the Colonial Museum, Wellington, gives the following measurements, which correspond very nearly with the size of the English species: length through the axis 3 inches 1½ lines, with a breadth of 1 inch 10 lines. White in colour, with the rough chalky surface which distinguishes the eggs of the Pelicanidæ.